THE CENTIPEDES AND MILLIPEDES (

CW01500936

R.F.LAWRENCE

# THE CENTIPEDES
# AND
# MILLIPEDES
# OF SOUTHERN AFRICA
## A GUIDE

1984
CAPE TOWN  A.A.BALKEMA  ROTTERDAM

# CONTENTS

# ILLUSTRATIONS

Figures in line drawings are placed with the relevant text.
Photographic illustrations: 2, 9a, 13a/b, 14, 15, 17, 22a/b, 24 a/b, 26, 30a/b, 33, 36a, 40a have been grouped together as a special section between pages 52 and 53.

# FOREWORD

*"Not a form so grotesque, so savage, nor so beautiful but is an expression of some property inherent in man the observer, — an occult relation between the very scorpions and man. I feel the centipede in me, — cayman, carp, eagle, and fox. I am moved by strange sympathies; I say continually 'I will be a naturalist'."*

Ralph Waldo Emerson. Journals III, p.163.

The Myriapoda of Southern Africa have received very little attention at the hands of zoologists in general and this small book is presented in the hope that, while it may be of interest to the reading public, it will also serve as an inducement to further studies by workers on the group. Though it deals in a general way with the centipedes and millipedes of Southern Africa, it is intended to serve more as an introduction to the fauna while providing as much essential information as can be contained within the covers of a small book.

Our knowledge of the South African fauna must rest mainly on a single volume, *The Myriapoda of South Africa* by Graf Carl von Attems of Vienna. It provides a framework and starting point for all South African students of the group. Although this volume, by a leading world authority, is invaluable and quite indispensable to any student of the myriapod class, it is almost exclusively a systematic work describing the nature and extent of the native fauna as a whole. To a certain degree it lacks completeness since Attems' monograph was based on the collections of the South African Museum at Cape Town alone. It reveals a faunistic bias since these collections were far more representative of the Cape province, especially the western half of it, than of the other provinces. The collections in the Transvaal Museum, the Albany Museum and the Natal Museum at Pietermaritzburg, were not consulted; the last omission was especially unfortunate as the Natal-Zululand fauna has proved to be richer in species than any of the other three provinces. Since the appearance of Attems' monograph in 1928 the number of species in southern Africa has almost trebled from 265 to more than 700 known forms.

The work done at the Natal Museum from 1935 to 1955 revealed the great number and rich variety of the myriapoda inhabiting the splendid forests of Natal and Zululand. All the indigenous forests of any size from Maputaland in the north-east to Port Edward in the south-west and from the coastal bush of the southern dunes to the montane forests of the Drakensberg kloofs in the north were visited, and collections made in all of them.

## Locomotion in Myriapoda
The most striking feature of the Myriapoda as a group is of course the multiplicity of legs, a condition which has been conveniently used when devising both scientific and common names as hundred- and thousand-legs.

Humans have two walking legs, most of the remaining vertebrates four; the insects can get along with six while the arachnids must make do with eight and most crustacea with ten. Only the centipedes, millipedes and *Peripatus* have a much larger number which in the millipedes may be more than 300.

How does the fleet-footed house centipede, *Scutigera,* manage to manipulate its 30 long slender legs when running at a speed of 25 m or nearly a mile an hour without putting a foot wrong? In Ray Lankester's little verse the confused centipede forgot 'which leg moved after which' and there must certainly be problems in placing the legs correctly. In the case of *Scutigera* especially there would have to be very precise stepping to avoid treading on its own toes.

In this important matter the outstanding authority of our time is Dr Sidnie M.Manton FRS who might almost be said to have perfected a new direction of research, a system which might perhaps be called a study of podologistics.

In her book, *The Arthropoda,* she explains how the various groups of arthropods have different rhythms and 'gaits' when walking which can change with an increase in speed. With such increases in the case of a centipede the animal would resort to a *modus operandi* which could be described as a 'gear change'. In *Peripatus,* also an animal with numerous legs, Dr Manton was able to distinguish a bottom gear used for starting, a middle gear for easy walking, and a top gear when the animal is doing the best it can. Fast moving centipedes such as *Scutigera,* which hunt in the open, would have to be quick off the mark and so lose no time in getting into top gear. The various gait patterns could be easily recognised and were characteristic of a particular order or group of myriapods.

Dr Manton also devised most ingenious ways of measuring the pulling powers of millipedes. When harnessing them to small trays or little pans to which weights were added, she found that they 'made most willing carthorses' and that their pulling powers, apart from their pushing ones, were considerable, surpassing those of a small mouse of a comparable mass. She even attached little boots to the feet of some of her millipedes to show up their gaits and to measure their stride lengths, the distance between foot-falls, as their speed increased.

Apart from the observational and experimental sides of her work, the exploration of the detailed anatomy of these small animals demanded infinite and meticulous care. Imagine for instance dissecting one of the legs of the *Scutigera* centipede, itself only about an inch long, and pin-pointing the 34 tiny muscles in it. Dr Manton's master-piece, *The Arthropoda,* published in 1977, contains many such drawings summing up the work of at least 20 years: as she describes it the book is about the comparative study of functional morphology and habits because habits and structure have evolved together. The habits and morphology of an animal are intimately tied together; there is no structure so bizarre or apparently useless but it has a meaning and a use and this can be discovered from the way it functions. By precise anatomical investigation and experiment on the living animal, by observing its habits and behaviour the two aspects can be correlated. Only on this basis can one proceed beyond speculation as to the origin and evolution of the various classes and orders of Arthropoda, their relationships to each other and the evolutionary conclusions to be drawn from such relationships.

Living animals for her researches were sent to Dr Manton from many parts of the world, some of them from South African forests. The temperate rain or mist belt forests of South Africa extend as a discontinuous belt along the coastal strip from the Cape Peninsula to Zululand. As I have shown in my *Cryptic fauna of forests,* they contain a unique collection of invertebrate animals, the great majority being arthropoda, embracing at least 32 orders; all the recognised orders with one or two exceptions are represented. They are thus as rich, if not richer, in the variety of their animal life than temperate forests anywhere in the world.

As if this were not enough nature has provided a further bonanza in the shape of *Peripatus,* a relic animal of the utmost zoological interest and importance, eagerly sought after by evolutionary biologists the world over. It is an animal limited to the tropics and to the temperate regions of the southern hemisphere. It is found in all our indigenous forests.

These forests thus represent living laboratories of richly variegated animal life, and they are situated at the very door steps of our universities. The three universities of the south-western Cape are served by the forests at Swellendam and Knysna, those at Port Elizabeth, Grahamstown and Fort Hare by the Hogsback and Amatola forests, while Natal and Zululand have at their disposal numerous low-lying forests as well as the montane ones of the Drakensberg.

The destruction of our beautiful and irreplaceable rain forests by the felling of these slow growing indigenous hard woods promises to be one of the greatest man-made disasters than can befall any land. The tragedy will be played out in the next 20 years and it is extremely doubtful if any unspoilt indigenous forest will be in existence by the year 2000. Once gone no amount of replanting or natural regeneration will restore them to their former glory.

Our indigenous rain forests are of such inestimable value and importance that we should be loath to sacrifice a single tree, yet the powers that rule our destinies persist in driving their four-lane motor highways through them. How many of the 25 million inhabitants of our land will be a penny the worse for not having such highways is a moot point; the developers, explorers and speculators will be the gainers and then the lesser parasites which follow along their trail. Esau has sold his birthright for a mess of pottage.

It is a relief to turn to those kind persons who by their publications or active participation have provided the assistance on which I have leaned heavily in compiling this book.

# INTRODUCTION

## THE MANY-LEGGED CREEPERS OR MYRIAPODS

The centipedes and millipedes are among the less well-known of the lower groups of animals, not only in South Africa but in nearly all countries of the world. The reason that they have attracted so little attention, as compared with the insects for instance, seems to be threefold. They exercise very little influence on human affairs, being neither of direct use to man like the silkworm and honeybee nor of interest to him as a source of food. In the second place their habits have none of the constructive qualities which so often engage our sympathy and interest in insects or in creatures like spiders. Thirdly, very few of them trespass upon man's preserves as pests, injuring his food crops or his personal possessions; none of them are known to cause or spread disease.

In spite of this universal neglect the centipedes and millipedes were recognised to be separate groups as early as the 4th century BC, since Aristotle, one of the great biologists of all time, and the first to devise a system of animal classification, had already distinguished the centipedes under the group name of Scolopendra) from the millipedes, Joulos or Julus .

In the *Historia Animalium,* one of his four main zoological treatises, he evidently recognised the essential differences which separate the group of Chilopoda (centipedes) from the Diplopoda (millipedes) as these groups are understood by modern zoologists. Aristotle remarked that they had no wings, designating them as 'long and many-footed insects'; he also noted that 'they live a long time after being cut in two, both halves of the body moving after the cutting, the legs near the cut end walking as well as those at the tail end'.

Most of the other men of ancient science fell very far below the high standard of scientific integrity set by their great predecessor. Among these was the Roman naturalist Pliny, credulous and uncritical in the extreme. This well-born gentleman and retired civil servant was apt to swallow the yarns of sailors and superstitious farm labourers whole, reproducing them in his *Natural History* which was widely read in the ages that followed. One such tale is that the inhabitants of Rhoeteum (a town in Asia Minor) were driven from hearth and home by a swarm of Scolopendra.

Aelian, about 200 AD, makes a much more matter of fact observation of the centipede Scolopendra when he writes, 'in the South there is a sort (of creature), from one to eight inches long, whose bite smarts like the sting of a wasp'.

If, as some authorities suppose, the Hebrew word which is translated 'mole', is really 'Scolopendra', this would be one of the earliest mentions of the centipede, and the only one in the Bible. The reference occurs in Leviticus: 30 and is part of the catalogue of 'creeping things that creep upon the earth' which are unclean as food and forbidden to the children of Israel. The list includes 'the ferret, and the chameleon, and the lizard, and the snail, and the ·mole'.

Linnaeus, the great Swedish biologist, who in 1760 constructed the first workable system of classification for animals and plants, and gave to so many of them their Greek and Latin names, regarded the centipedes as wingless insects; in this he was perhaps not so far wrong as might be supposed. The name Myriapoda was originally used in 1796 by Latreille to embrace the centipedes and millipedes which with certain Crustacea he ranked as an order under the insects.

As far as is known the earliest description of a South African centipede in scientific literature occurs in the *Codex witsenii,* a travel record which embodies a number of paintings and descriptions of animals and plants collected by Simon van der Stel, during an expedition to Namaqualand in 1685-6. The two coloured portraits of a centipede which it contains were made at Cape Town in 1692 by Hendrik Claudius; though crude, they are clearly recognisable as the large *Scolopendra morsitans,* a species common throughout Southern Africa; the artist has erred in only one respect, drawing the creature with 19 pairs of legs, thus giving it two pairs less than the rightful number.

In Europe the first recognisable drawings of millipedes known to us have an earlier origin, and some were executed by Ulysses Aldrovandi in 1602.

The modern names of millipedes or thousand-legs, and centipedes or hundred-legs, are much the same in most European languages, and we have Millepied and Centepied in French, Tausendfüssler and Hundertfüssler in German, names which express the same idea and which would occur quite naturally to people of different languages and cultures. Although there are not by any means as many as a hundred legs in the centipede or a thousand legs in the millipede, these names are good and useful ones which do bring out the great difference between the number of legs in the two groups.

In other languages also, centipedes and millipedes are called by names which derive from the number of their legs and in some cases they imply considerable powers of accurate observation. Sinclair in the *Cambridge Natural History* tells us that the Arab name for centipede in the Middle East is 'arba wak arbarin' or 'forty-four legs'. This is almost the exact number possessed by the larger centipedes in hot countries, such as Scolopendra, which actually has 21 pairs of legs, or, if the large leg-like poison claws at the anterior end of the body are included, 44 in all.

The Persians have more generalised names for them, 'Hazarpa' or thousand-feet, being equivalent to our Millipede, 'Sadpa' or hundred-feet to our Centipede. In African languages the Swahili term for centipede is 'Tandu', for millipede 'Jongvoo', the Zulu equivalents being 'Ikume' and 'Songololo', the latter being the colloquial name usually adopted by South Africans to denote millipedes or myriapods in general.

The name 'Songololo', though used indiscriminately for all Myriapoda by both English and Afrikaans speakers, should apply to millipedes only since the body structure of the centipede differs widely from that of a millipede and it cannot enroll in a flat spiral. The word is derived from the Xhosa verb Ukusonga, 'to roll up' and according to Dr Jean Branford, in her *A dictionary of South African English,* Xhosa women call the coiled intra-uterine contraceptive device 'Songololo'.

Primitive peoples as a rule have extremely acute powers of observation with regard to the living creatures around them, distinguishing even the lower orders of animal life as individuals by singling out some peculiarity of appearance or habit, and crystallizing it in a word. It is not surprising then that even the Australian aborigines have a name for the centipede, which they call 'Thingathinga Ya'.

## ACKNOWLEDGEMENTS

The inspiration derived from the work of the late Dr Sidnie Manton I have already referred to. I also owe much to that of Dr Barbara Brunhuber now lecturer at the University of Ontario, Canada. Little has been done on the biology of the Chilopoda in South Africa so that the original contributions of new knowledge embodied in her studies on the reproduction of the Cape Centipede, are a refreshing change that should act as a spur to the present generation of students of the Myriapoda.

My sincerest thanks are due to my colleague Mr Peter Croeser who has always been available with helpful comment and constructive criticism and who furthermore has been burdened with the task of reading through the manuscript.

Lastly let me express my gratitude to Mrs Huibré Tomlinson for providing an excellent typescript and for dealing so efficiently and speedily with endless corrections, second-thoughts and emendations.

R.F.Lawrence

# I. SOME GENERAL OBSERVATIONS ON THE MYRIAPODA

Many people take centipedes and millipedes to be the same kind of animal without making any sort of distinction between them. Worse still, others lump them together with the smaller legless wriggling creatures as simply 'worms'. They are of course quite different from the lower orders of real worms, such as earthworms, tape-worms and the bristly sea-worms which we find living among rocks on the sea-shore; these all have soft bodies and no jointed legs, a fact which can easily be ascertained by examining the ordinary earthworms of our gardens. Nor do the centipedes and millipedes have any direct relationship with the assortment of garden pests usually called by such names as army-worm, cut-worm and boll-worm, which are no more worthy of the name of worm than is the thousand-legs. These all represent the larvae or caterpillar stages of insects, usually moths, beetles or flies, and though worm-like in outward appearance, are very different both in their internal and external structure.

The centipedes and millipedes are for the sake of convenience bracketed together by zoologists in a single group called the Myriapoda or many-legged creatures. The likeness between them however is more apparent than real, as the millipedes differ in many important respects from the centipedes. This general similarity is partly due to 'convergence', by which a superficial likeness between unrelated groups of animals is brought about owing to both groups being subjected to the same conditions of life. Both millipedes and centipedes have become suited for movement on or under the surface of the ground and frequently have to accommodate themselves to rather cramped living quarters under stones and logs. They are adept at disappearing into small holes or crevices in the soil, for which their long and rather narrow bodies are admirably suited. In the same way the porpoise and the shark rather resemble each other because a streamlined shape suits both of them for travelling rapidly through water, though the first is a mammal and the second a fish.

On account of their living in dark, out of the way corners under stones and fallen tree trunks, if not actually under the surface of the ground itself, the sense of sight in both centipedes and millipedes is poor. In one large section of millipedes and also in one of the four centipede orders, all the members are blind, while in other sporadic cases the eyes may be either absent or greatly reduced. In only one of the various myriapod groups has keen eyesight been developed, the fleet-footed pouncing scutigeromorph centipedes whose predacious habits will be described hereafter.

1

## 1. *The differences between millipedes and centipedes*

Let us consider a few of the differences which distinguish millipedes from centipedes. In the first place their reactions to an irritating stimulus, such as being touched by a piece of stick, are quite different. When disturbed the millipede at once coils its body round its head in the shape of a clock spring, a move which protects not only the head but the soft under belly of the animal; if irritated further by rough handling it may secrete an evil smelling substance from special glands or void liquid faeces.

The centipede however has a very different response for it immediately becomes galvanised into action and darts away with all the speed of which its many legs are capable. The coiled up millipede if it is left in peace for a few moments will begin to unroll itself, cautiously at first, and then with its body fully extended will move off, walking, or rather gliding along in a more or less straight line like a small railway train. In the movements of the centipede there is no straightforward gliding motion but, as it gets up speed, an S-like undulation such as seen in the progression of a snake; this movement is too fast to allow us to make out the details with the naked eye but if motion pictures of a running centipede are taken it can be seen that the legs of each pair on opposite sides, except those at the front and hind ends of the body, are moving in different directions, while the left hand leg of a pair may be directed forwards the right hand leg is pointed backwards (Figure 1). In the millipede on the other hand the opposite legs of each pair are moving together in perfect unison and can be compared with the rowing action of a man sculling a boat; if we watch the millipede from the side we see what appears to be waves moving along the rows of legs from behind forwards like those made by the wind over a cornfield; about five of these rhythmic or metachronal waves can be counted at any one time caused by the momentary lifting upwards and forwards of groups of legs, each group consisting of five or six pairs

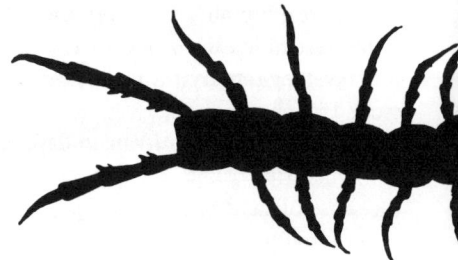

(Figure 1). All such peculiarities of walking and other functions in centipedes and millipedes may not seem very important but actually they are an indication of the very different way in which the parts of the body are constructed; such differences of arrangement are again associated with great dissimilarities in the habits and functioning of the two groups.

A few of these differences in the makeup can be seen if we examine specimens, preferably dead ones, of these animals. Let us take as an example of a centipede the large green-banded *Scolopendra*, and of the millipede the large brown or black spirostreptid thousand-legs often seen on the open veld and which is perhaps better known by the Zulu name of 'Songololo'. We notice first that the millipede has two pairs of legs to every ring or segment of the body, from which is derived the scientific name of the class, Diplopoda or double-footed, each ring segment or diplosomite, which really consists of two fused segments, also carries two pairs of tracheal systems opening on the sternal plate near the bases of the legs; but the centipede on the other hand has but one pair of legs to each body segment though these are stronger than those of the millipede which seem small and weak when compared with its bulky and heavy body. Secondly the body of the millipede is encompassed by an unbendable skeleton composed largely of calcium while that of the centipede is fairly flexible since it is enclosed by a tough but fairly thin covering of chitin similar to that which we find in insects; the centipedes are characterised by their flexibility, being, in contrast to the incompressible millipedes, able to flatten their bodies and in some cases to shorten or extend them, enabling them to pursue prey or find a safe refuge in such confined situations as crevices of rock, cracks in dry sun-hardened soil or under the bark of trees. Thirdly the body of the millipede is most often round and cylindrical or hemispherical in cross section while that of the centipede is flattened dorso-ventrally to a greater or lesser degree, as in some insects. Fourthly the millipedes are with few exceptions vegetarians, feeding on large

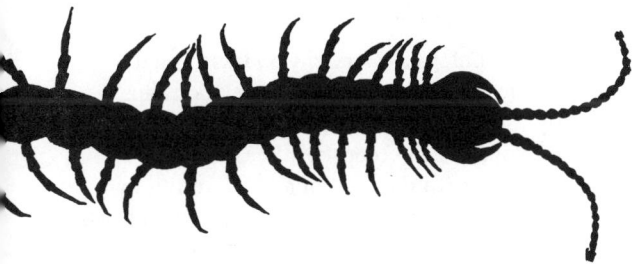

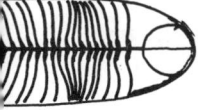

Figure 1. The walking movements of a centipede, left, and a millipede, right, seen from below.

3

quantities of decaying wood or leaves, such as is found littering the floors of forests, and occasionally on living plants while the centipedes are all flesh-eaters and armed with poison claws by means of which they paralyse and capture their prey. Lastly we come to a very important difference though doubtless of more concern to the specialist and student than to the educated layman interested in the life around him. The millipedes are distinguished by a peculiarity of structure found in comparatively few other animals, namely the openings of the sex organs being situated near the front of the body, not far behind the head; the centipedes on the other hand resemble the insects in this respect where these openings occur at the extreme hind end of the body. These are only a few of the more obvious differences between the two classes of animals; there are of course many more refinements and differences in detail with regard to structure, habits and life histories but these can safely be left to the specialist whose concern they are.

Even if centipedes and millipedes were not so obviously different from each other one could say that in centipedes the number of antennal joints is large and varies from 14 in *Geophilus* to over 400 in the house centipede, *Scutigera* (Figure 17). All the millipedes on the other hand, even those which hardly look like millipedes, such as the tiny 'pin-cushion' millipede, *Polyxenus,* have short antennae with the same number, neither more nor less, of eight segments.

Another more homespun difference is that millipedes cannot swim but sink like stones when dropped in water; their drowned corpses can often be found in gutters or street run-offs after flooding by heavy rain. Centipedes on the other hand, being able to flex their bodies sideways, can swim by undulations as snakes do in water. This must play a significant part in the distribution of the two groups, for when large rivers form the natural boundaries of a given region, this disability will have a more restricting influence in the case of millipedes.

## 2. *The place of the Myriapoda in the animal kingdom*

Where do the Myriapoda stand in the zoological hierarchy and with what other main groups of invertebrates do they have their closest relationships? For many years and until recently they have been simply regarded as a subdivision of the vast phylum Arthropoda which comprises the Crustacea, Arachnida, Myriapoda and insects, each with the rank of a Class. In this matter, the writer shares the more recent views on classification of such zoologists as S.M.Manton and O.W.Tiegs and which have been clearly set out by Dr Manton in her book *The Arthropoda* of 1977. She maintains that there are such deep-seated differences between the members of the old system of the Arthropoda that the rank of phylum should be assigned to each, at the same time regrouping them under three heads, Chelicerata, Crustacea and Unira-

4

mia, with the extinct primordial Trilobites as a possible fourth phylum.

The Chelicerate phylum comprises the extinct scorpion-like Merostomata, their still living relative the king-crab *Limulus,* and the terrestrial and carnivorous Arachnida. The second phylum, the Crustacea, is an aquatic one. The third, the Uniramia is the phylum with which we are concerned in this book; it comprises three terrestrial groups, the Onychophora (*Peripatus*), the Myriapoda and the Hexapoda (insects in the wider sense). All three have been promoted and become sub-phyla. The Myriapoda in this arrangement stand closest in their development and structure to the Onychophora and Insects, having little relationships with the Arachnida and hardly any with the Crustacea.

We have now arrived at the subphylum Myriapoda itself which is divided into four classes. Two of them, the well-known Chilopoda and Diplopoda, which between them make up almost the entire fauna, are redivided into ten orders. The remaining two, the Symphyla and Pauropoda, are far less familiar; when compared with the centipedes and millipedes they are primitive and small, both in numbers and body size; like *Peripatus* they are compact and homogeneous groups, with only a few forms which can be accommodated in a single order each.

Altogether the status of these arthropod phyla and their relationships to each other present an extremely complex problem with room for a number of divergent opinions. Nevertheless, it is interesting to note that Prof Adam Sedgwick, the distinguished Cambridge zoologist, seems almost 100 years ago to have arrived at conclusions not very different from those of such modern authorities as Manton and Tiegs.

In a slightly different way he expressed the relationship simply in the following words:

The classes Insecta, Onychophora and Myriapoda are the survivals of a once great and continuous group of land Arthropods, a large number of which have become extinct, leaving two groups, Insecta and Onychophora, each fairly compact and showing little variety of organisation, and one, the Myriapoda, loose and heterogeneous, with considerable gaps between the orders.

The place of the Myriapoda in the recent systematic arrangement of the Arthropoda according to Manton (1977) is shown in the Table below.

SUPERPHYLUM ARTHROPODA

| Phylum | Subphylum | Class | Order |
|---|---|---|---|
| 1. Chelicerata | Merostomata | Three | |
| | Arachnida | | many |
| 2. Trilobita | | | |
| 3. Crustacea | | many | many |
| 4. Uniramia | Hexapoda | many | many |
| | Onychophora | one | one |
| | **Myriapoda** | Chilopoda | 1. Geophilomorpha |
| | | | 2. Scolopendromorpha |
| | | | 3. Lithobiomorpha |
| | | | 4. Scutigeromorpha |
| | | Diplopoda | 1. Pselaphognatha |
| | | | 2. Oniscomorpha |
| | | | 3. Juliformia |
| | | | 4. Polydesmoidea |
| | | | 5. Nematophora* |
| | | | 6. Colobognatha |
| | | Symphyla | one order |
| | | Pauropoda | one order |

* The only Myriapod order not to occur in South Africa; for the most part it inhabits the countries of the Northern Hemisphere but two or three genera are found in tropical Africa.

# II. THE CENTIPEDES IN GENERAL

Before describing the particular kinds of centipedes in greater detail it would perhaps assist the reader if an account were to be given of the characteristics, structure and habits of the centipedes in general. This class in zoological terminology bears the name of Chilopoda, the Greek word meaning 'thousand-feet'.

## 1. *The situations in which they are usually found*

Centipedes are not commonly met with in our walks over the veld as they are retiring creatures and take refuge during the daylight hours under stones in open country or under fallen logs in our indigenous forests; we shall therefore have to go and look for them in these places if we wish to make their close acquaintance. A few of the larger centipedes like *Scolopendra* provide exceptions to this rule and are sometimes found walking in the open especially on misty or cloudy days when the direct rays of sunlight are not so potent (Figure 2); they will also be found more often towards the evening or the early morning when the air is still damp from the dew. The explanation for the fact that centipedes choose sheltered places to live in can be found in their breathing organs or tracheae which are not as highly developed as those of the insects; they do not have such efficient control over the intake or release of air from these tubes so that they are liable to loose moisture from their bodies in this way; also the outer skeleton of chitin, is not as thick and rigid, especially at the sides of the animal, as that of the insects and is less impervious to the sun's rays; these two weaknesses mean that under a hot sun they are liable to become dessicated and expire within a short time. They thus spend the daylight hours in crannies where the air contains a certain amount of moisture and venture forth in the cool of the night to hunt for food and mates.

Some of the Geophilomorpha of northern lands seem to be able to live along the sea-board of rocky coasts where they prey on small marine crustacea and worms and are able to withstand immersions by the sea at high tides; no semi-marine forms however seem to live along the South African coasts; true cave-dwellers (troglobionts) are found among the medium-sized lithobiids, though not in large numbers. With the exception of the blind genus, *Cryptops,* the larger Scolopendromorpha do not seem able to tolerate the cold and humidity of underground caverns.

## 2. *The colouring of centipedes*

In the colours of their bodies the centipedes are no match for the insects with

7

their brilliant iridescent colours. The great majority of species favour drab colours, usually yellow or brown, and in many cases there is only one colour without contrasts or pattern markings to relieve the monotony of the general effect. An exception however must be made in the case of some of the larger centipedes of which *Scolopendra morsitans* is a good example. The typical form of this species is yellow with dark green cross bands on each segment while the race of *morsitans* found in Namaqualand has a dark indigo-blue body and legs a vivid orange, quite different from the typical banded form. In the forest-living *Cormocephalus anceps* the head, first segment and last two or three legs are a bright red contrasting with the rest of the body, the antennae and legs being blue-black. Both the leaf-footed centipede *Alipes crotalus* of Natal and the widespread and robust *Cormocephalus nitidus* are a uniform terra-cotta or brick-red colour, though the smaller adolescent stages may be light green.

### 3. *The organs of respiration*

The breathing openings or spiracles of the centipede are of the same nature as those of the insect, and like them are placed at the side of the body where they can be seen between the upper and lower chitinous shields and just behind the legs (Figure 3). These small openings of the breathing tubes when examined with a magnifying glass look like little buttonholes and are easy to see as they are much darker than the membranous skin or pleurites, at the sides of the body. They are either round, triangular, or S-shaped, and each leads into a large wide tracheal chamber which branches off to supply the various organs within the body, the legs, and the antennae, with oxygen from the air. The greater activity of some insects requires a pumping mechanism to accelerate the flow of air through the spiracles and we see this at work in the rapid and rhythmic pulsations of the body of a grasshopper when it is held in the hand. There is no such arrangement in the centipede but its wriggling movements serve two purposes, for not only do they enable it to get a better grip of the ground and thus to move faster, but they also bring about a better circulation of air in the tracheae. In general however the amount of air required by centipedes is small and they can survive immersion in water for a day or longer. In most centipedes the spiracles are placed on alternate segments of the body, and in the large *Scolopendra* for instance they are to be seen on segments 4, 6, 8, 11, 13, 15, 17, 19 and 21.

### 4. *The sense organs*

If we further examine the large *Scolopendra* centipede we find that the head and mouth-parts are more or less flat, enabling it to enter restricted and narrow crevices and even to seize and consume its prey there. At the front of

8

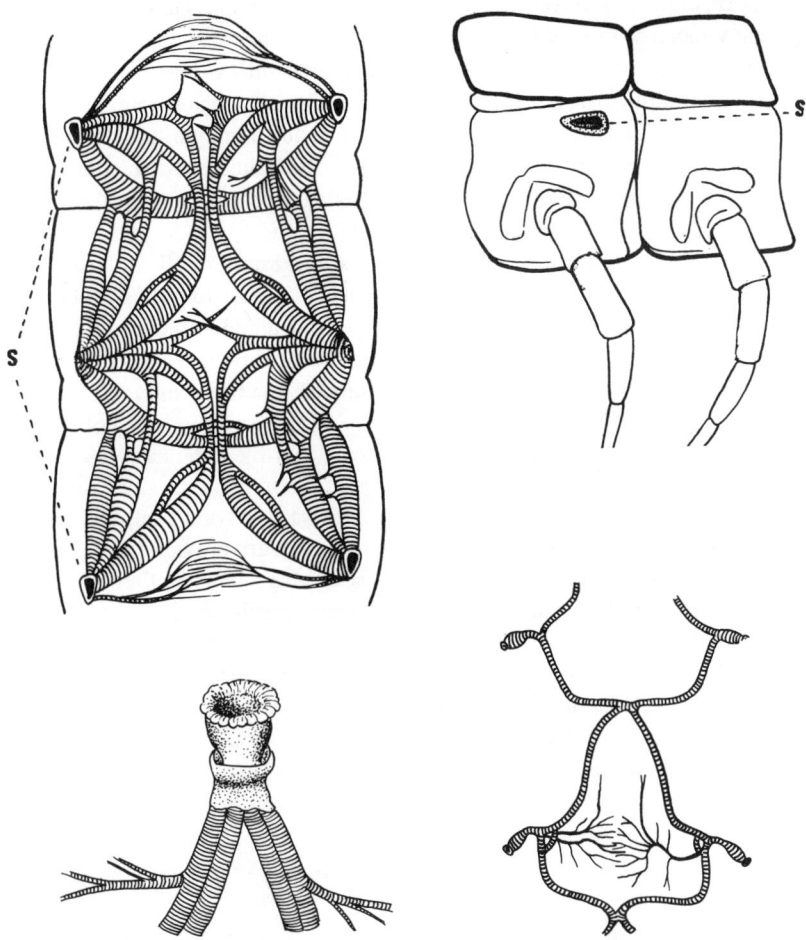

Figure 3. Above, left, three segments of the centipede, *Scolopendra* showing the tracheal system; right, one of the triangular spracles (*s*). Below, left, the tracheal system in two segments of *Geophilus;* right, the round spiracle enlarged, with the main tracheal trunks attached.

the head are the long forward pointing antennae, each with 17 joints giving them great flexibility so that they can be bent and twisted to a quite remarkable degree (Figure 5a). These antennae seem to be the most restless parts of a centipede and are used almost like a whiplash for they keep up a constant flickering movement while the creature is in motion, tapping over the ground and exploring it as a blind man does with his stick. This is not surprising as eyesight is very poor in the centipedes and there are usually only four minute single eyes at each side of the head, probably little more than organs for distinguishing

9

darkness from light. The four eyes or ocelli are of the simple kind, not the large many-facetted eyes of insects but of a type which is also characteristic of the Arachnida such as spiders and scorpions, each with a single lens. Most centipedes certainly do not depend on their eyes for running down and securing prey or for avoiding their enemies and quite a number have no eyes at all. An exception must be made in the case of the 'house-centipede' *Scutigera* which has large compound or facetted eyes of which more will be said in the section dealing with this order of centipedes.

The antennae on the other hand are the most important of all the structures which carry organs of sense as they are richly provided with hairs and bristles which are extremely sensitive not only to solid and liquid objects but to currents of air and probably also to changes of temperature. Nearly all these tactile hairs are concentrated on the antennae, the body and legs being comparatively smooth and thus less sensitive to contact with external objects; it is of course fitting that this should be so seeing that the antennae would be the first part of the moving centipede to come into contact with any living or non-living objects. In centipedes which walk backwards a good deal, such as the geophilids or earth centipedes, the last pair of legs are provided with many sensitive hairs and when the centipede is moving in reverse they act as feelers, taking no part in walking. On the antennae, usually at their extreme tips, there are a number of rod-like or peg-like chemo-receptor organs which correspond to the sense of smell, or perhaps a combination of smell and taste (Figure 4). It should be noted also how closely the sense of touch and smell are allied, so much so that usually no awareness of smell can take place unless the object in question is first touched.

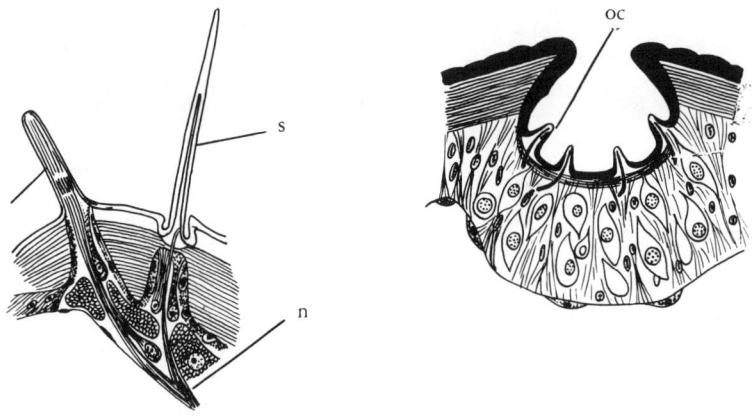

Figure 4. Left, a cone-shaped olfactory organ (*oc*) with nerve branch on the antenna of the stone-centipede, *Lithobius*. Right, a pit-like organ of smell on the antenna of *Scutigera* with four olfactory sense cones (*oc*).

10

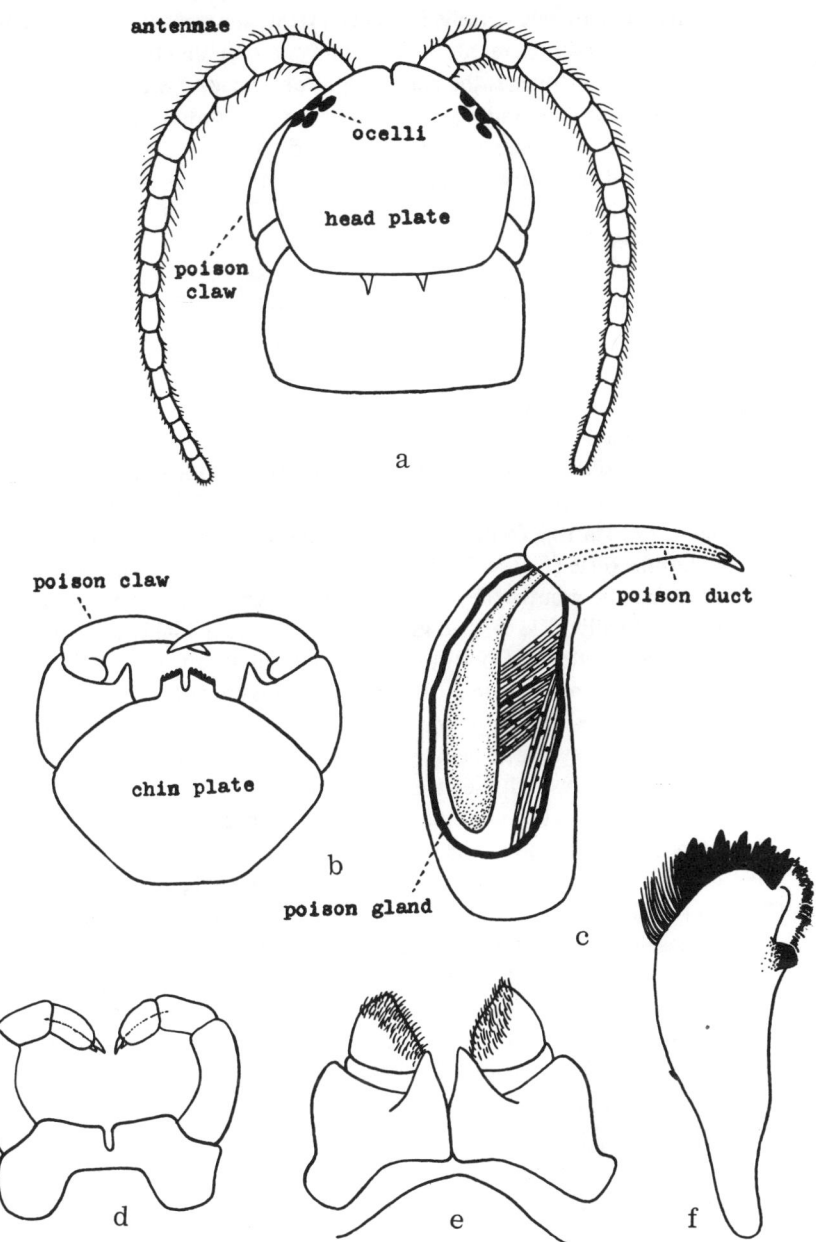

Figure 5. Mouth-parts of *Scolopendra:* a. head from above; b. chin-plate (gnathochilarium) and poison claws; c. a poison claw enlarged showing the poison gland and its duct; d. first maxillae; e. second maxillae; f. mandible.

To watch a centipede confined in a glass jar with a cricket or cockroach is to realise how blind the creature is. It blunders rather aimlessly round the jar until by accident it touches the prey with its antennae; it is then galvanised into immediate activity and makes a number of clumsy lunges in the approximate direction of the insect; in one or other of these lunges it makes contact and the victim is then very quickly grasped and bitten with the large poison claws.

## 5. *The mouth parts with which the prey is captured and eaten*

The mouthparts with which the prey is captured, cut up, and ground small correspond piece for piece with those of insects and are very much alike throughout the various groups with the possible exception of the geophilomorph centipedes.

As all centipedes are carnivorous there is a pair of strong mandibles or jaws provided with powerful teeth and spines for masticating the food (Figure 5f); following these starting from above come two pairs of appendages, the first and second maxillae (Figure 5d, e), which play a role in holding the food and guiding it into the mouth; they are also provided with a taste sense which however can hardly be as refined as the sense of taste in higher animals. Below these maxillae are the large and formidable poison claws which are the main organs of offense and defense in the centipede (Figure 5c). It is difficult to know whether to call these 'poison claws' or 'poison jaws' because though they are jaws in the sense of being very close to the mouth, which they assist in the holding of the food, they were originally the first pair of legs of the centipede which have become enlarged and developed as a poison weapon. The sharp fang at the end of the organ corresponds to an ordinary claw of the leg which has been perforated by a fine canal opening near its sharp tip. This duct leads down to the large poison gland which is buried deep in the muscles near the base of the appendage.

When we turn the centipede over on its back we will see first the large and conspicuous chitinous plate which forms the floor or underside of the head, the syncoxite, comprising the two fused coxal segments (Figure 5b), to its sides are attached the large poison claws or toxicognaths. To examine the other chewing parts of the mouth, the poison claws and large plate must be lifted with a sharp scalpel; when this has been removed the following parts will be uncovered: first the pair of second maxillae, below these the pair of first maxillae and finally underlying all and touching what might be called the 'roof' of the mouth, the paired mandibles. (Fig.5, p.11)

Last, but not least in importance, four large glands, in two pairs, open into the mouth or buccal cavity; they provide salivary and digestive secretions for breaking down the raw animal protein of the prey. In feeding the poison claw is used to make an initial opening in the cuticle of prey which has already

12

been immobilised; the tooth-like cusps on the inner side near the base of the claw can then be used as a kind of 'tin-opener' by a sudden jab to enlarge the hole that has been made by the claw; the weaker and more delicate mouth parts, the first and second maxillae, and the mandibles with their powerful teeth are used to tear out the soft parts of the prey assisted by digestive juices poured from the mouth into the wound. If the prey is large, such as a soft-bodied gecko, the whole head of the centipede may be inserted into the fissure which has thus been made. In most centipede groups the first leg is larger and stronger than the rest and takes a part in the feeding process by helping to hold the food mass steady.

It is interesting to compare these mouthparts with those of arachnids such as the harvest-spiders and scorpions. The Arachnida have no antennae or mandibles; the part played by the last named structure is taken over by the basal parts of the first and second pairs of legs which are arranged in a rough circle round the opening of the mouth and assist in manipulating the food; this state of affairs represents the early crude attempts in the direction of proper chewing mouthparts, and in those ancient ancestors of the arthropoda, the trilobites, the bases of the first four pairs of legs were often armed with teeth and took the place of jaws in chewing the worms and other small animals on which they fed.

The centipedes thus show a distinct advance upon the arachnids (spiders and scorpions) in having proper chewing tools such as the mandibles and maxillae; in most insects these have been still further improved, while the crayfish has a whole battery of gadgets round the mouth for dealing with its food.

## 6. *The segments of the body*

Behind the head of the centipede come the large number of body segments, which are practically all alike; each segment is built up of a roof-shield or tergite, a floor-shield or sternite, and on each side, a side-shield or pleurite. The shields covering the top and bottom of each segment are of chitin and these, while strong, are flexible so as to allow for a large amount of the turning movements which are so necessary in the running activities of the centipede.

Flexibility of the body is also gained by having these tergites of different lengths, a long plate alternating as a rule with a shorter one as is well seen in the lithobiomorph and scutigeromorph centipedes. In the large robust Scolopendromorpha such differences are slight and in the worm-like Geophilomorpha the numerous segments are of uniform length but are provided with short intermediary segments between the normal ones. A contraction of the body with the tergal shield telescoping by sliding backwards over the one behind it, also provides greater flexibility.

The pleurites or side shields, on the other hand, have very little chitin but are soft, membranous, and almost white in colour. On these can be seen the openings of the tracheae, the spiracles or stigmata (Fig.3, p.9)

13

## 7. *The legs and their peculiarities*

One of the obvious things about a centipede is of course its large number of legs; passing backwards along the body it can be seen that these legs, all except one pair, are alike and are usually composed of six joints ending in a stout claw which rests on the ground when the creature is walking, or enables it to grip the surface when climbing fairly steep slopes. The last pair of legs, the end-legs or anal legs, are rather different from the others; in the first place they are longer and thicker and are usually armed with a number of very strong, prickly spines (Figure 6a). They can be used for grappling the prey or another belligerent centipede, in pincer-like fashion, in which case the numerous spines help to hold it securely. It has been noticed that some centipedes when attacked lift up their end-legs and brandish them in the face of the aggressor. One of the large centipedes found in Natal has been named *Alipes* or wing-footed on account of the extraordinary shape of the end-legs which are so flattened and thin that they resemble small leaves or wings (Figure 6c). While this centipede is walking, but more especially when it is irritated, these leaf-like legs are waved to and fro, fanning the air with a fluttering motion and giving rise to a faint rustling sound. It is perhaps a method of inspiring fear into would-be aggressors and in this resembles the stridulation or hissing noises made by scorpions and sunspiders (Solifuges).

Another peculiar and interesting arrangement is found in the end-legs of another small scolopendromorph, *Cryptops,* a blind centipede, where these legs are covered with sensitive tactile hairs while on the under side of three of the segments there is a row of saw-like teeth. The end segment hinges back on the one behind it like the blade of a pen-knife on its handle; any small insect that unwittingly touches one of the sensitive hairs causes the end segment to snap back against the neighbouring one so that it may be trapped between the two segments and impaled or held fast by the sharp teeth of the under side of the leg (Figure 6b).

The end-legs of one peculiar species of centipede, *Cormocephalus cupipes,* living in the forests of Natal and Zululand, are broad and unusually flattened above. It seems to live exclusively under the bark of trees and this habit of life is probably responsible for the modification which has taken place in the form of the end-legs. A similar flattening of the whole body is found in many insects and spiders which make a permanent use of such cramped quarters.

## 8. *The manner of walking*

The act of walking in centipedes, as in other animals with many legs, opens up àll sorts of interesting problems, some of them biological, other mechanical, which would require a whole chapter to themselves. The locomotory mechanisms of the Myriapoda and other arthropoda have been dealt with in great detail by Dr S.M.Manton in her book *The Arthropoda.*

14

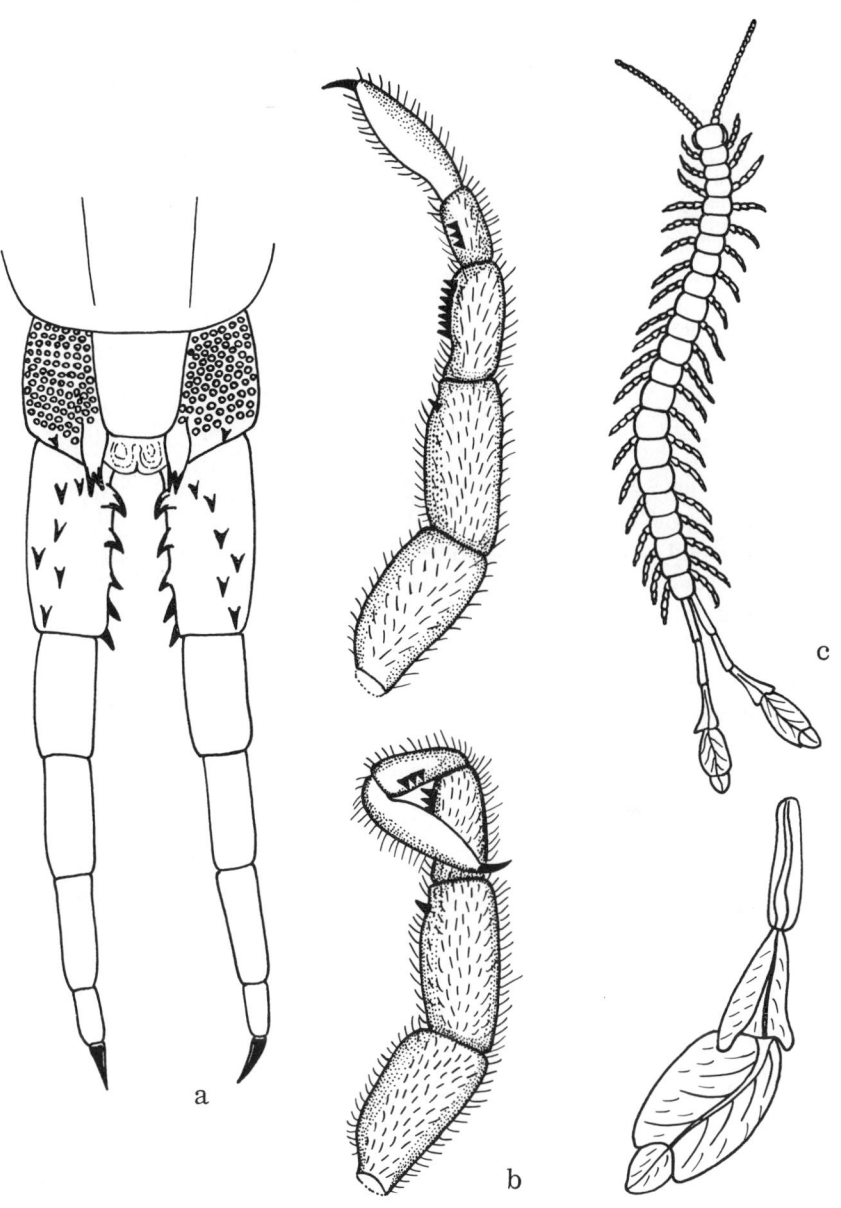

Figure 6. a. the end-legs of *Scolopendra* seen from below; b. end-legs of the blind centipede, *Cryptops*, with the tarsal segments extended and closed; c. the wing-footed centipede, *Alipes*, with the wing-like end-legs enlarged.

15

As has been already mentioned, the gaits of centipedes and millipedes are very different; in the rather clumsy and slow-moving millipedes the legs are weaker than those of centipedes but while these small and feeble appendages hardly suffice to lift the millipede off the ground, the work of moving it is shared by a larger number of participating members so that many legs make light work. The legs of each pair in the millipede move simultaneously or in synchronism while in most centipedes they move alternately. To use the analogy of rowing the legs of the millipede have a sculling action while those of centipedes can be compared to two rowers seated side by side each using his oar alternately.

Speed of movement is not dependent on the number of legs in the many-legged arthropoda, or the millipedes with their larger number of legs would move faster than the centipedes, instead of the reverse being the case. Speed is rather a function of the length of the leg and its mechanical strength and efficiency. That increased speed in centipedes is partly obtained by an increase in the muscles of the legs is shown conclusively by Dr Manton's anatomical investigations. She was able to identify 13 muscles in the legs of Geophilo-morph centipedes, 19 in the Scolopendromorpha, in the Lithobiomorpha 20, while in the fleet-footed Scutigeromorpha there are no less than 34 different muscles for each leg. This in contrast to the slow-moving Diplopoda where the number is only two or four.

In centipedes, those with the largest number of legs are also the slowest moving, since these many legs are small and weak, while those with the small-est number, which are also long and strongly built, are the fastest. If the four large divisions of centipedes, represented by well-known types, were to be arranged to form a series in which the number of legs decreases but the length of the leg increases progressively, they would stand in the following order: *Geophilus, Scolopendra, Lithobius* and *Scutigera.* In *Geophilus* there are numerous legs, all short and weak, but an advantage for a centipede of burrow-ing habits; *Scolopendra* has a fair number of longer and stronger legs, *Litho-bius* fewer but still longer legs, while *Scutigera* has the same number as *Litho-bius* but by far the longest legs of them all. The speed of movement in the four groups of centipedes can also be arranged in this order, *Geophilus* with its short and feeble but numerous legs, in some cases as many as 170 pairs, being much the slowest mover, *Scutigera* with its 15 long pairs by far the fastest. Examples of the three first named groups when tested for speed showed 43 cm per minute for *Geophilus,* 300 for *Scolopendra,* 525 for *Litho-bius* while the running speed of *Scutigera* was measured at 2 500 a minute or a little less than a mile an hour.

The worm-like *Geophilus* assists the legs by sinuous movements of its body and since its short legs hardly lift it off the ground it has to curve itself round small obstacles which it may meet in its path. The long legs of *Scuti-gera* on the other hand lift its body well off the ground and over such obsta-

16

cles, so that its course is much more direct. It does not employ undulatory movements to aid its locomotion and in contrast to the worm-like *Geophilus* its construction and appearance is more like that of an insect.

The ability to walk or run backwards is most highly developed in the many-legged centipede *Geophilus* which is able to do this as easily as running forwards; the other groups can only retreat backwards for short stretches and when obstructed they will more often turn completely round, if there is room for such a manoeuvre, instead of reversing in their tracks.

When centipedes are cut into sections the pieces still continue to move and the legs to perform walking movements, as was commented on by Aristotle in the first zoological text-book written 2 000 years ago. Such walking movements are thrown less out of gear in the groups with many legs than in those with few legs and the separate parts of a centipede such as *Geophilus* will still behave in a fairly normal manner; small pieces of this centipede, consisting of only two or three segments are able to walk in a reverse as well as a forward direction.

In the same way locomotion is less disturbed by removing some of the legs in those centipedes with a large number of appendages, such as the *Geophilus* and *Scolopendra* groups, especially if the first five pairs behind the head are left intact. In *Lithobius* the co-ordination of the legs in walking is upset only when more than three consecutive segments in the middle of the body are deprived of their legs.

When the lower part of the brain, the suboesophageal ganglion, is removed, *Geophilus* is not much affected and moves normally while the other groups move at half speed or cannot move at all. If half the brain is removed centipedes move in a circle towards the side on which the brain is still intact, but at a reduced speed.

## 9. *The simplicity of the habits of centipedes*

The centipedes have no complicated habits such as are characteristic of groups like the insects and spiders. Their manner of life is eminently simple in the sense that they make no elaborate nests or retreats, neither do they dig burrows like the trap-door and wolf-spiders. They have no camouflage designs of colour or form for concealing themselves from their enemies or from their own potential prey. Still less do they make the complicated geometric snares which the sedentary spiders spin for trapping other small creatures. They can be imaged as free-booters and bandits living off the land, primitive in their few and simple needs, killing and eating what they require for food on the spot, and for a retreat taking advantage of any suitable stone or crevice which will protect them from the extremes of heat and cold and the ravages of storm and flood. The nearest approach they make to constructing any sort of shelter is the very roughly scooped out hollow in soft decaying wood which the female prepares when the time has come for her to lay her eggs; even this is only made by some of them, and many centipedes unconcernedly let their eggs fall to the ground, leaving them to their fate.

The sex and mating habits of centipedes, which are described more fully under the sections dealing with the separate orders, are fairly simple and rudimentary. There is a certain amount of sexual interplay in the form of repeated tactile stimulation, which may be mutual, of the antennae and end-legs but bodily contacts are very limited and of short duration, with the possible exception of the Scutigeromorpha. Thus the preliminary courtship of the male is brief and without the elaborate dancing displays and postures before the female which are a feature of many of the Arachnida, such as scorpions, false-scorpions and whip-scorpions (Pedipalps). There is no copulation. Like other carnivores the centipedes have developed no vestiges of community life with its grades and castes, and its marked division of labour.

## 10. *What centipedes feed on*

Although centipedes can go without food for several months before suffering any ill effects, in captivity they feed readily and thrive on a diet of mealworms (the larvae of the flour beetle *Tenebrio molitor*). If a mealworm is introduced into the living quarters of a starved scolopendrid, the centipede will not recognise it by sight as prey but only when the mealworm has been touched by the antennae. It will then lunge forward onto the worm, holding it with the anterior pairs of legs, and biting it several times with the poison claws. It then immediately begins to feed at one of the areas which has been bitten and where the cuticle of the worm has presumably been ruptured; feeding commences before the worm has died and leaves behind only the empty shell of the cuticle and sometimes not even that.

A mealworm held with the forceps and moved to and fro near the head of the centipede will bring no response. If a starved centipede has been unduly disturbed by some break in its routine it may run straight over the prey without seizing it and afterwards make no attempt to find it again. On other occasions after a disturbance it will bite the prey rapidly with the poison claws and then toss it away in the same way that a puff adder tosses a rat, leaving it to die from the effects of the poison bite.

With the exception of the Scutigeromorpha, centipedes are creatures which avoid light, hunting and feeding for the most part at night. The smaller species prey on small soft-bodied insects, worms and spiders, of which there are great numbers in the dark and damp localities where they live. Geophilid centipedes, in keeping with their subterranean environment, may subsist almost exclusively on earthworms in regions where these are abundant. The largest centipedes, those belonging to the genus *Scolopendra,* which may attain a length of over 10 inches in the tropics, have been known to attack lizards, frogs, and even small birds. One of these large centipedes, which was kept for some time in capitivity at the Zoological Gardens in London was fed on young mice. In South West Africa the writer has seen small geckos bitten in the neck by the large *Scolopendra morsitans* succumb very quickly, and doubtless these lizards were sometimes caught and eaten by them. In some parts of the world they are a deadly menace to trapdoor spiders whose doors they force open and then pursue the unfortunate householders down their own burrows, where, unless there is an emergency exist as a second line of defence, they are invariably cornered and devoured.

## 11. *Animals which prey on centipedes*

Centipedes are blood-thirsty creatures and will fight and destroy others of their own kind if they should inadvertently meet with them. They have however few special enemies but are occasionally eaten by birds and more often by small burrowing animals like moles and rats which may happen upon them in the course of their tunnelling activities. Some of the primitive Ponerine ants, themselves half-blind creatures living a largely subterranean existence, hunt the smaller centipedes for food. Probably some of the larger carnivorous slugs such as *Apera,* indigenous in South Africa, include geophilid centipedes in their diet. The subterranean-living slug *Testacella,* which also occurs in South Africa, appears to live on nothing else in certain regions of the world.

As far as I am aware only one animal, a snake, feeds mainly on centipedes. This is the centipede-eating *Aparallactus capensis,* a pretty little cream-coloured snake with a black collar band behind the head which can often be found in soil near the surface of the ground or under stones but is also fond of living in termite nests where the young are able to feed largely on termites. It hunts the larger kinds of centipedes such as *Scolopendra* and *Ethmostigmus,* for food. The habit is more pronounced in related species of this snake, living in tropical East Africa where centipedes are extremely plentiful.

Centipedes thus have few specific enemies and in general when hard pressed are able to disappear quickly down a convenient alley way in their semi-underground habitat or conceal themselves by squeezing their bodies into a narrow crevice where they are more likely to escape notice; in general also their inconspicuous colouring which makes them difficult to distinguish against a rough and dull-coloured background of soil or dry wood, is in their favour; the dim twilight and half light which reigns perpetually on the floor of dense forests also tends to blur outlines which would be sharply defined in the clear light of day.

## 12. *The poison of centipedes*

Little is known about the chemical composition of the poison which is acid and has a haemolytic action resembling that of some spiders and scorpions. Vertebrates and most arthropods (spiders, insects, crustacea) are very sensitive to the poison, while fishes, worms, and mollusca are more or less immune to its effects.

The large elongate poison glands in the Scolopendromorpha are buried deep in the tissues of the poison claw coxa but not far behind the fang; in the Geophilomorpha however they may lie well behind the head as far back as the 12th-18th body segments, the poison being conducted to the fang by long narrow ducts which open a little before the fang-tip.

Though painful, the bite of the centipede seldom proves fatal to man and in most cases is probably not much worse than a bee sting. In more severe cases it is said to resemble the thrust of a red hot needle in the flesh giving rise to considerable swelling in the neighbourhood of the bite; there are intermittent shooting pains through the limb affected; the action of the heart is fast and irregular and the patient suffers from cold sweats accompanied by feelings of anxiety and oppression, symptoms resembling those which result from the bite of the 'black widow' spider *Latrodectus*.

In the great majority of cases however a complete recovery is made within a few days. This is not always the case in the tropics however for it has been reported that the giant centipede of the Solomon Islands, probably a species of *Scolopendra*, causes fatalities every year. It is understandable that persons who are undernourished or in delicate health, and especially young children, are liable in certain cases to succumb to the bites of the large tropical centipedes, but the smaller kinds, living in the more temperate parts of the world, such as those inhabiting Southern Africa, are practically harmless; these may merely prove a nuisance as they are apt to hide in clothes, shoes or household receptacles left lying about.

A curious case of death indirectly due to the bite of the large *Scolopendra* is related by the French writer Bachelier; it concerns an officer in the colonial army at Tongking who on a hot night drank in the dark from a carafe containing a centipede; the centipede bit him in the gullet causing such a swelling of the sensitive mucous lining of the throat that he died of asphyxia. One species of South African centipede, *Cormocephalus nitidus,* gives off an unpleasant foetid odour when handled or confined in a glass tube; it is not known how this effect is produced and in no other centipedes of South Africa has it been reported. It may possibly be a secretion from the numerous glands which open on the coxopleurite of the end-legs in all Scolopendromorpha; no known function has been ascribed to these glands with their numerous pore-like openings though Dr Barbara Brunhuber thinks that the secretions may aid the sexes in finding each other.

Several cases have been reported to the writer of the rash-like effects produced by the mere walking of a large centipede over the more sensitive parts of the hand or arms; in all these cases the passage of the centipede raised a red inflamed weal where its legs had touched the skin; accounts of such happenings, from several parts of the world, seem too similar in their nature to be ascribed to mere superstitious fancy.

In a letter to *Nature* some years ago Mr R.A.Lever mentioned that in the Solomon Islands a Scolopendra centipede had one evening walked over his hand, which was touched by about 20 of the legs; as a consequence he experienced a numbness in the back of the hand which lasted the rest of the evening and throughout the night. It is difficult to give a simple explanation of such occurrences seeing that there are no known poisonous substances directly connected with the legs of any centipede, which are in every way similar to those of other creatures such as spiders and insects; as has already been pointed out, only the large poison jaws near the mouth have poison glands embedded in them. It is possible that the psychologist may provide an explanation, especially if there is an abnormally intense fear of centipedes in the mind of the person concerned in the incident; another would occur to the zoologist who might point out that centipedes frequently and with great care, clean their bodies, including all the legs, which they pass one after the other through the mouth; there is little doubt that they then become covered with saliva with which may be mingled a certain amount of venom from the toxicognaths or poison jaws. There at any rate we must leave the matter for the present.

## 13. *Losing legs to save lives*

One curious method of self-protection which is developed in many centipedes must certainly play a part in assisting them to escape from what enemies they have.

The habit of throwing off a limb which has been seized by some other predatory animal is a practice which is widespread among many of the lower forms of life and everyone is familiar with the way in which some lizards, such as geckos, when seized by the tail can readily part with it, leaving the appendage in the hands of the would-be captor and making off with only a stump. The tail of such a lizard fractures at a weakened point and for some time after breaking from the body seems to possess an independent life of its own, wriggling and twitching in a lively and rather disconcerting manner. Though a tail is lost, freedom and perhaps life itself may be gained, since a new tail is soon grown, and nobody is any the worse.

This habit of casting off a tail or a leg, known by such names as autotomy or autospasy, is widespread in many of the invertebrate animals, such as the arthropods with their jointed limbs, and fragile worms such as the common earthworm and the flatworms. It is practised by a number of the Arachnida, such as spiders, Solifuges, harvest-spiders and mites, especially by those which have greatly elongated and slender legs. It is much less common among the insects, and some of the arachnid orders, such as scorpions, do not seem to practise it at all. Some of the long-legged harvest-spiders (Palpatores) on the other hand, have been known to lose all except two of their eight legs, and still manage to get along. Sphincter muscles round the joint where the fracture occurs close up the opening and prevent undue loss of blood.

The method in this apparent madness is simple; one of the joints of the legs which are liable to be discarded is weaker than the others, and under any strain the leg parts easily from the body at that point. The joint may be a different one in different animals; thus in spiders the break occurs between the first and second joints of the leg, the coxa and trochanter, in harvest-spiders between the second and third joints, in the whip-scorpions (Pedipalpi) between the fourth and fifth joints.

Among the Myriapoda, all the centipede orders have a number of representatives which are liable to lose their legs easily, while in the millipedes on the other hand it does not seem to be a special habit, seeing that all the legs in this group are rather delicate and are never very strongly attached to the body, so that any of them may become detached if specimens are roughly handled.

Casting off the legs reaches its greatest frequency in the long-legged house centipede, *Scutigera*. All the legs of this centipede are easily lost, but especially the last and longest pair, so that it is very difficult to procure a specimen for Museum collections with all the legs intact; when *Scutigera* is drowned in tap water at ordinary temperature, the last pair and a number of other legs are often thrown off. In the stone-centipedes, of which the best South African representative is the widespread *Paralamyctes,* with their shorter legs, autotomy is also quite common, the posterior legs again being the ones most liable to be lost, but on the whole the process is not carried to the extreme found in *Scutigera.* Few of the large centipedes such as *Scolopendra* and *Cormocephalus,* part with their legs readily, not even the last pair, but an exception must be made with regard to centipedes which have the last pair of legs strongly modified, as is the case in *Cryptops* and *Alipes.* In *Cryptops* where the last pair of legs have been transformed into trapping organs, large numbers of spirit specimens in Museum collections which have been killed by immersion in alcohol, have lost either one or both of the end-legs. In the large and robust leaf-footed centipede, *Alipes crotalus,* the pair of flattened and leaf-like legs at the end of the body come away very easily when grasped with a forceps, the break occurring between the coxa and trochanter, or the first and second joints. In the small and worm-like centipedes, *Geophilus* and its allies, with their extremely numerous short legs, it is usually only the last pair of transformed legs which is discarded.

All these legs after parting from their owners continue to twitch spasmodically for a short time, like the broken off tails of lizards, as if they had an independent existence of their own. As in nearly all other small arthropods the lost limbs of centipedes are soon regrown, and after one or two moults a new and perfectly formed leg appears to take the place of the old one which

22

has been sacrificed.

Two points of interest emerge from these various observations on the shedding of legs in centipedes. One is that it occurs more often and more readily in those with long and slender legs such as *Scutigera* and *Paralamyctes,* and secondly that it is more often the legs at the hind end of the body than those at the front that are lost. In the large centipedes, such as *Cryptops* and *Alipes,* where only one pair of legs is liable to be discarded, it is the last pair which is concerned. This is quite what we should expect to find if the habit is to be regarded as a means of self-preservation, seeing that a centipede in flight would present its rear end to a pursuer.

## 14. *The parasites of centipedes*

Centipedes are very clean creatures and, as will be described later, groom themselves scrupulously and frequently; they seem to be almost completely free from small external parasites such as the small red-bodied trombidiform mites which so often attach themselves temporarily to the bodies of spiders, harvest-spiders (opilionids) and occasionally, scorpions. One or two of the larger tropical scolopendrids harbour a species of mite which is a permanent lodger and runs actively about the body of the host; in such cases each species of centipede has its particular species of parasite found on no other centipede. None of the South African centipedes are known to harbour such mites though a number of our larger millipedes certainly do. Enormous numbers of the deutonymph or wandering stage of tyroglyphid mites living in forest humus attach themselves to nearly all cryptic arthropods including centipedes but only temporarily for transportation. They are liable to cluster in large numbers on or around the genitalia of both centipedes and millipedes but especially the latter.

Internal parasites living in the intestines of centipedes are also comparatively few in number and probably harmless; one or two protozoa, Microsporidia, Coccidia and Gregarina have been found in various centipedes; large numbers of the nematode worms that infest the gut of most cryptic arthropoda parasitise centipedes while they are also sporadically the hosts of Mermithid worms and parasitic flies.

Centipedes may be attacked by a number of bacteria and fungi. Moulds are the worst enemies of myriapoda in general, attacking the eggs and young more frequently than the adults.

## 15. *The dispersal of centipedes by commerce*

On account of their flattened bodies and habit of inserting themselves into narrow dark crevices, centipedes are ideally suited for being transported in cargoes of merchandise, such as crates of fruit, from one end of the earth to

23

the other. In Hamburg, one of the great sea ports of the world, Dr Karl Krae-pelin found that 490 species of animals from foreign lands were imported by shipping in three years; of these 28 were centipedes, but not one of them sur-vived the first winter in Hamburg, or spread further than the bales of mer-chandise lying on the quays. In this way a few species of centipedes have become naturalised in South Africa and have spread very rapidly throughout the country.

## 16. *Centipedes and man*

J.H.Fabre once wrote 'The world from the gastronomic point of view is a robber's cave and man the greatest bandit in it'; the most omnivorous of all animals, he has spared few of the creatures which share his planet but seems to have made an exception of the centipede. Few tribes or peoples however primitive, and however hard pressed by food scarcity seem able to overcome a natural repugnance for poisonous animals, such as centipedes and scorpions, so that they are seldom used for food. In Siam, it is true, centipedes are roasted and given to children suffering from 'thinness and swollen belly', and it is said that under the excitements of religious fanaticism African Arabs swallow them alive, together with prickly pear leaves, pieces of glass, and other un-pleasant objects. As a rule however these animals inspire a certain healthy prejudice in most people who are revolted by their all too many legs and wriggling movements, which are like those of snakes in miniature; to this is added a natural and genuine fear of their poisonous qualities. While conceding the fact that the centipede is a blood-thirsty and destructive creature with no beneficial or even useful qualities that we can discover, the astonishing care with which the mother centipede tends her young does to a certain extent bridge the gulf that separates us from such a totally different order of life.

# III. THE FOUR ORDERS OF CENTIPEDES

Almost 2 000 species of centipedes are found throughout the world and of these about 150 have their home in South Africa. They fall naturally into four large divisions or orders which are all well represented in our country. These four can be easily distinguished from each other by their outward appearance alone and are named after four type genera, *Geophilus, Scolopendra, Lithobius* and *Scutigera,* the Scolopendromorpha for instance being composed of all the large *Scolopendra*-like centipedes. All these genera are found in South Africa except *Geophilus* which however is represented by closely related forms.

The four groups will now be described in turn but the Scolopendromorpha are given fuller consideration than the other orders as being the centipedes most familiar to people in general, the one they would be more likely to have met with on excursions and walks. They represent more closely the norm or common idea of what people visualise as a centipede and are less specialized in their habits and structure, standing nearest to the generalised ancestor of the Chilopoda.

## A. THE GEOPHILOMORPHA OR EARTH-CENTIPEDES
Type: *Geophilus* Leach.

### 1. *General description*

The Greek name of this order of centipedes, *Geophilus* or earth-dweller, tells us in two words quite a lot about its habits. The Geophilomorpha are almost entirely subterranean in their manner of life, being very seldom seen on the surface of the ground or exposed in the open. They are good examples of animals whose structure and outward appearance have become changed in response to a very restricted and specialised habit of life. Many creatures which live in perpetual darkness, such as moles and burrowing lizards, have very small eyes or tend to lose them altogether and the *Geophilus* centipedes, which live in a similar environment, are all blind. These worm-like centipedes differ from all the other orders in having much more slender bodies which are usually very long and composed of numerous segments varying in number from 35 to 170. It is not easy at first to distinguish head from tail end of the creature as these are often very similar in appearance; the fact that it can move backwards almost as readily as it does forwards makes this even more difficult. In response to the cramped environment in which these centipedes live the legs, though numerous, are very short and rather weak. The last pair at the hind end of the body is a little longer than the others, resembling in appearance the antennae at the front of the head which are short in all Geophilomorpha with a constant number of 14 segments. These end-legs can in fact be used as substitutes for the antennae when the centipede walks backwards, as it often does; they are extremely sensitive, being covered like the antennae with large numbers of tactile hairs by means of which they are able to continually explore the ground. Males, which are otherwise similar to the females, can be distinguished by means of these legs which are often longer and sometimes considerably thicker than those of the females, or are provided with a fur-like covering of innumerable fine hairs. The basal (coxal) segment of these legs is often greatly swollen and is provided with a large porose area resembling a sieve, with numerous pores which are the openings of single-celled glands beneath the skin (Figure 7).

The Geophilomorpha are exceptional in having areas with numerous minute gland openings, the porose areas, on many of the ventral shields; the pores are arranged in groups, usually two or four in number. The colourless secretion of these glands has an acid reaction and soon coagulates in the air. Little is known about their function though they seem to play a role as a defence weapon (Figure 7).

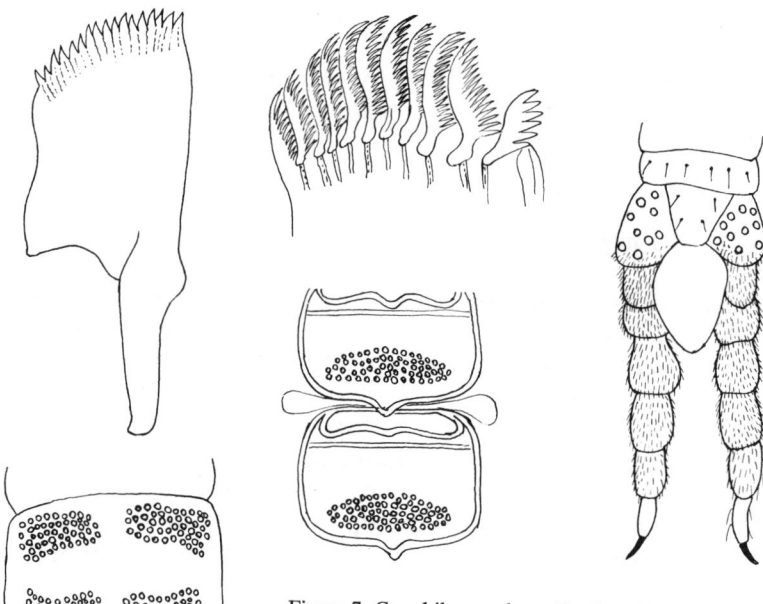

Figure 7. Geophilomorph centipedes. Above, the mandibles of two families. Below, the arrangement of the porose areas of the ventral shields in two families, and right, the end-legs of the South African geophilid, *Purcellinus*.

26

## 2. *The conditions under which they live*

The earth-centipedes live largely in loose soil or damp mould and gardening enthusiasts are very liable to turn up specimens with their forks when excavating compost heaps. Short spineless legs which do not get in the way and a long body are advantages for an animal which is above all a burrower; the longer the body the more deeply the centipede can burrow. The shape of each segment can be altered, becoming elongate or short and thick so that an individual can readily change its length or width. Greater flexibility of the body is obtained by the addition of short intercalary or intermediate segments inserted between both the dorsal and ventral shields, the tergites and sternites.

The manner of burrowing adopted by the Geophilomorph centipedes, as observed by Manton, can be briefly described as follows: the anterior segments and head which are a little narrower than the following segments are pushed wedge-like into a crevice in the ground while the first pair of legs, which is thicker and stronger than the others, anchors the animal by holding fast to the ground gained; the segments behind shorten and thicken, exerting pressure and widening the aperture already made; the segments in front then elongate and resume the advance, pushing further into the crack, while the body segments from behind again join up with them. The process is repeated intermittently, heaving against the soil until the centipede can pass through the gap (Figure 8).

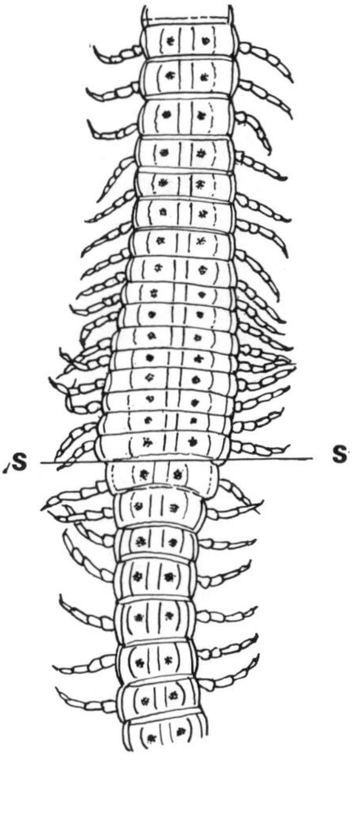

Figure 8. A geophilomorph centipede accumulating pressure to get past an obstruction, a glass plate $S - S^1$ by shortening and widening the body segments (after S.Manton).

27

Although comparatively slow-moving the Geophilomorphs are usually extremely difficult to catch as they take advantage of the smallest crack in which to disappear; before they can be grasped with a forceps they are already some inches under ground. When running the very supple body can be bent sideways in strong undulatory movements but if cornered they may roll up into a loosely knotted ball, like an untidy clew of string, and lie quiet. Another device practised by some of these centipedes in an emergency or escape situation is to adopt the 'looping' method of progression characteristic of the caterpillars of geometrid moths. The body is raised in a loop by alternately anchoring a few of the anterior and posterior legs to the ground.

The mouth-parts of the Geophilids are small and delicate in comparison with those of the other three centipede orders; the mandibles lack the strong tubercular teeth which other centipedes use for grinding up the prey; they are replaced by a series of flexible, weakly chitinised plates with rows of long slender comb-like teeth which seem better adapted for straining semi-solid food than for masticating large parts of the prey.

The Geophilomorpha are the most difficult of all the centipedes to identify as in each case the minute mouth-parts have to be dissected out and mounted for microscopical examination; in many Geophilomorpha there has been considerable reduction in some of the appendages of the mouth-parts which often appear rudimentary. In keeping with these generally fragile and poorly developed mouth-parts, the long oesophagus is extremely narrow, almost thread-like, leading into a stomach which is only a little wider. It must be presumed that in this order of centipedes the food actually swallowed cannot be of a solid nature but has first to be liquified by means of digestive secretions. With their weak mandibles they are usually only able to feed on small soft-bodied insects and worms of which there are vast numbers in the rich soil that Geophilids favour as a place to live in. In Europe they hunt and attack earthworms and some species seem to live entirely upon them; they have been seen to attack earthworms much longer and thicker than themselves, coiling round them like pythons strangling their prey and biting them to death with their small poison claws.

### 3. *Light producing centipedes*

A curious habit which distinguishes these small centipedes from the other orders is their ability in some cases to give out a light resembling the 'phosphorescence' of the glow-worm; although this luminescence has as yet been observed only once in the case of South African geophilids it is very probable that others of the order can produce it.

The light-giving substance is a sticky colourless liquid secreted by glands which open through numerous small pores on the ventral shields of the body and which at certain times can become luminescent; its production is con-

28

Figure 9. An old illustration of a European light-producing centipede (photo R.A.Holliday).

sidered to be a defence reaction and when the centipede is attacked or irritated by enemies such as ants, its body lights up with this greenish fire (Figure 9). It can also be produced by artificial stimulation such as slight pressure or dropping the centipede into water. The light is powerful enough to be seen at a distance of 9 m (30 feet) and by its aid print can be read 10 cm away; it is produced with equal intensity by both sexes while in the glowworm that of the female is the more brilliant; it occurs in centipedes of various ages. It is not known what benefit the production of light can be to its possessor; it has been suggested that it is a means of enabling the sexes to find each other but seeing that all the earth centipedes lack eyes, a light-giving organ would not be of much benefit to them in this respect. It may on the other hand have some value as a protection for the animal, perhaps by keeping the body free from harmful bacteria and fungus spores that live in soil and are liable to adhere to the skin; this is perhaps the more likely explanation seeing that these centipedes, on account of their small mouth-parts,

29

are not able to brush and polish their bodies and legs in the way that the large Scolopendrid centipedes do, though the antennae are regularly and carefully cleaned.

In South Africa luminescence as far as I am aware has been observed in only one centipede, *Orphnaeus brevilabiatus;* the species, seen near Salisbury, is widespread in tropical and subtropical regions of the world and well-known for its luminescing proclivities.

### 4. *The situations in which they are found*

It is unlikely that the reader will have seen an example of these small and rather insignificant centipedes except perhaps in the exhibition cases of Museums. They are not common and never occur in large numbers, each individual living more or less by itself. In compost or manure heaps and in the soil of well-tended gardens they are more numerous, as these situations contain a rich supply of the small animals which serve as food for them. The gardener digging in a part of the garden where the soil is particularly rich in decaying substances may turn one out with his fork. They are found everywhere throughout the world and are the only centipedes which can to a certain extent be called marine in habit as some live on the sea shore in crevices of the rocks where they are able to withstand many hours of submersion when the rising tide covers their retreats.

### 5. *Their relations to man*

These little creatures are neither useful nor harmful to man and their poison claws are much too small and weak to injure any but the smallest insects. In some very rare instances they have been held responsible for cases of false parasitism by creeping into the nose or ear cavities of sleeping people; in so doing they set up violent irritation, resulting sometimes in vomiting, hysteria and paroxysms; such mishaps are however entirely accidental and extremely rare. For the rest the earth centipedes merely play their obscure role in underground darkness; in some way as yet not sufficiently understood they have their place as a link in the vast and complicated network of living things of which man himself forms a part.

### 6. *Mating, egg-laying and development*

Of the four orders of centipedes the simplest forms of association between the sexes occurs in the Geophilomorpha. According to Klingel there is a somewhat brief preliminary courtship in which the female is prepared for receiving the spermatophore by means of simple tactile stimuli such as tapping and stroking with the antennae, more or less as in the larger Scolopendromorph centipedes; after some mutual tapping of the antennae and posterior legs, the partners leave each other. The male then spins some threads across the tunnel which they are at the time occupying and deposits a droplet of sperm, without a covering envelope, in the zig-zag web; after an interval the female moves forward along the burrow moving the posterior end of the body rhythmically in a 'wiping' motion; once the web is found and contact made it moves over it with strong wiping movements of the posterior end and picks up the sperm droplet, depositing it in the genital opening. The threads spun by the male thus discharge a two-fold function, providing a platform for the naked sperm droplet and at the same time serving as guide lines to the female for finding the spermatophore.

The earth centipedes bring forth and care for their young in much the same way as do the Scolopendra centipedes. The eggs differ from those of their larger relations however in being round instead of oval (Figure 9); as would be expected they are much smaller and more numerous, any number up to 65 being laid in the spring or early summer. There are three main stages of development, in the last of which the young can be recognised as centipedes, though they are still very small and quite white. The mother shows the same care for her young as in the *Scolopendra,* guarding them without leaving the brood chamber or taking food during the critical and dangerous eight weeks when the helpless young are growing up.


## B. THE SCOLOPENDROMORPHA OR LARGE CENTIPEDES
Type: *Scolopendra* Linnaeus


### 1. *General description*

The Scolopendromorph order is represented by the type *Scolopendra,* a Greek word meaning 'Palisade-bearer'. The name, one of the oldest in zoological history, was conferred by Aristotle on all the centipedes and probably refers to the row of legs on each side of the body. The Scolopendras now include only the large centipedes, which are well represented in South Africa, some of our species being at least 7 inches in length.

The centipedes of the Scolopendra group are in the main inhabitants of the tropics and subtropical lands; very few are found in the colder northern latitudes. The largest species in the world, *Scolopendra gigantea,* is found in Columbia and various islands off the coast of Venezuela; this colossus among centipedes may measure nearly 11 inches (26,1 cm) in length and almost half an inch (1,25 cm) thick. Nearly all the large forms like *Scolopendra* and *Ethmostigmus* live in India, Malaya, Africa and the tropical belt of South America.

Our own country has plenty of representatives, though they are somewhat smaller than the giants of the tropics; most of the centipedes in fact which people come across in the open, belong to this group, as the smaller ones described in subsequent chapters are often too insignificant in size to be noticed, or live in surroundings which are unfamiliar to most people.

These centipedes can be easily distinguished from any other by their large strong bodies which consist of about 25 similar segments, and by their considerably enlarged end-legs which are armed with conspicuous spines (Figure 6a). The antennae have 17-20 segments but usually 17. They have large poison claws and nearly all the cases of bites giving rise to serious consequences in human beings have been due to members of this group. They have a habit of slipping into shoes or clothes which have been left lying about overnight, and people have been bitten on their hands and feet in this way on a number of occasions; such bites are said to be rather common in the Hawaiin islands of the Pacific.

The Scolopendras are among the few centipedes which venture out into the open, but even they are very seldom seen in strong sunlight, preferring a misty or a cloudy day for their walks abroad, or the hours near sunset and sunrise. Their movements are only moderately fast but when pursued and running at full speed are very characteristic, being serpentine or S-like, resembling those of snakes, or lizards with very short legs. Some of these larger centipedes are extraordinarily restless and one species, *Cryptops hortensis,* kept under observation at night in red light, has been seen to keep on the move for more than an hour without stopping for a single moment; the flickering antennae hardly ever ceased to move.

## 2. Colouring

A bright colouring of the body, with patterns and markings, is exceptional in centipedes generally, but is found to a greater extent in the *Scolopendra* group than in any other centipedes. *Arthrorhabdus formosus,* a very common centipede throughout the Western Cape and midlands, is coloured deep blue, with the head and last segment a rich purplish red. *Scolopendra morsitans,* which is wide-spread throughout the continent of Africa with the exception of a small corner, the Cape Peninsula and its surroundings, has a yellow body with dark green cross bands, the head, antennae and last two segments being blackish, and the legs bright orange, a very handsome and striking uniform. In the open country we are sure to meet with both these two species, which usually live under stones. In forests and indigenous bush one is more likely to come across a smaller bronze-green centipede under old logs or in the damp humus of the forest. This is *Cormocephalus,* found almost everywhere in South Africa and represented by a large number of species (Figure 2); in the drier parts of the country, where forest is absent it is well able to live under stones. In our forests another interesting centipede is found, though it is often over-looked on account of its small size and dull brown or yellowish colouring; this is the blind centipede, *Cryptops,* which is almost exclusively an inhabitant of forests and is seldom found in open country under stones. The leaf-footed *Alipes* (Figure 6c), with its peculiar flattened end-legs, also belongs to the group of large centipedes and lives in Natal, Zululand and the northeastern parts of the Transvaal. The entire creature, body, antennae and legs, is coloured a light and rather attractive orange or terra-cotta red.

### 3. *Toilet or grooming habits*

Centipedes spend a lot of time over their daily toilet; they clean themselves with great care and in fastidious detail, reminding one very much of a cat licking itself all over with calm deliberation as if that were the only really important business of the day. As we watch a captive centipede hurrying along in that restless and apparently aimless manner which is so characteristic of its tribe, it is more than likely to stop suddenly and in a quite inconsequential manner begin to clean its antennae; rather as the absent-minded professor pauses to polish his spectacles in the street. The antennae are drawn sideways through the mouth while the jaws appear to nibble at it; this is done carefully several times until the individual segments are not only cleaned but coated with saliva from the mouth. It can of course be understood how important it is for the centipede that these parts of the body should be kept scrupulously clean seeing that they carry the most valuable of all its sense organs. After taking a meal the antennae are usually cleaned again and the centipede may then begin with the toilet of the body and legs. Each leg is held by the large poison claw and passed through the mouth from base to tip; the centipede goes to work systematically, first down one side of the body and then down the other. After this the body shields are cleaned and in this the small leg-like second maxillae of the mouth play the most important part, passing over them with a circular polishing movement so that minute specks of dust and dirt are brushed off. The head and segments just behind it are cleaned by passing a number of the legs behind the poison claw over them; in much the same way as a cat smooths the back of its head with its paws.

### 4. *The daily round*

Dr Barbara Brunhuber has given us a full and factual account of the day to day activities, or perhaps it would be better to say, the nightly activities, of captive males and females kept in a vivarium at the University of Cape Town where, after ascertaining the exact degree of humidity necessary for their well-being, they lived for some years, feeding and thriving on a diet of mealworms. Her specimens all belonged to the widespread forest-living species *Cormocephalus anceps* which like centipedes in general is nocturnal in habit; moribund during the day, their activities begin at dusk and close by about midnight. When a number of both sexes are kept in an enclosed space there are bound to be chance encounters and sooner or later many of the smaller juveniles became the prey of the larger cannibalistic animals. Centipedes are not immune to their own poison and in the unnatural and more crowded conditions of captivity many of the weaker succumbed to bites from the poison claws of stronger individuals.

Centipedes are ruthless and pugnacious creatures but sometimes when two males of approximately equal size meet, a fight to the death is avoided by both going into a defensive posture, the two contestants lying side by side with the enlarged spiny end-legs of each firmly gripping the body of the other behind the head so as to forestall an attack, rather like the locking hold used by wrestlers to immobilise an opponent (Figure 10).

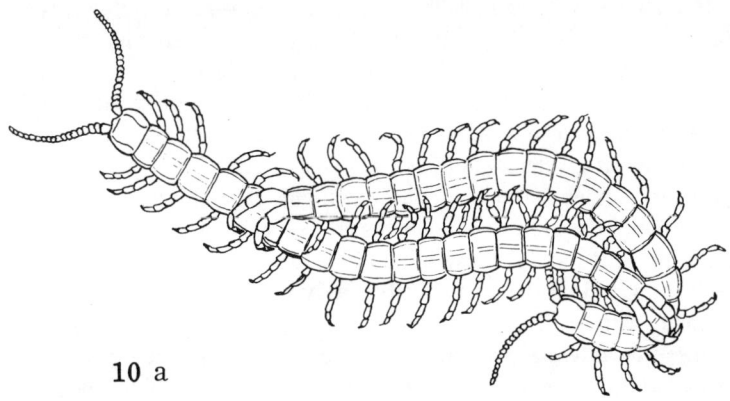

10 a

In encounters between the sexes it was usually the male that was killed and eaten, often while still alive, by the more aggressive female, even when she was the smaller of the two. Such behaviour is no doubt very similar in other species of *Cormocephalus*, *Scolopendra* and probably all the large centipedes of the Scolopendromorph group. I well remember on one occasion in the field turning over a large stone and finding a pair of large *Scolopendra* centipedes under it. The male presented a grisly spectable as the female had bitten off his head and was busily devouring her decapitated partner with her head almost buried in his still writhing body.

### 5. Courtship and mating

The sexes do not differ in size or colour and secondary sex characters have been recorded in the males of only one scolopendromorph, *Otostigmus*, a genus not found in southern Africa. The sexes can therefore only be distinguished by differences of the genitalia which are invisible for most of the time. The genito-anal complex can be extroverted by using gentle pressure on the penultimate segment of the ventral surface in suitably narcotised speci-

34

mens. Living animals can be almost completely immobilised by keeping them for some time in a refrigerator. Specimens which have become hardened in the liquid preservatives of museum collections should be held for some time with the posterior segments immersed in hot water.

Mating, in the sense of contact between the sexes by means of copulation does not apply to the centipedes as there is no intromittent organ in the male. Courtship, though it takes place, is of a more simple and rudimentary nature than that practised by many other arthropods such as the spiders, scorpions, false-scorpions and pedipalpi, in which the female is conducted through a kind of ritual dance, consisting of a number of steps or figures, or the male executes these in front of the female.

Mating behaviour in any degree of completeness has only been described twice. The first account is by Klingel in 1960 for *Scolopendra;* the second six years later and the only one of a South African animal, is that of Barbara Brunhuber in 1967 which differs in a number of respects from Klingel's version.

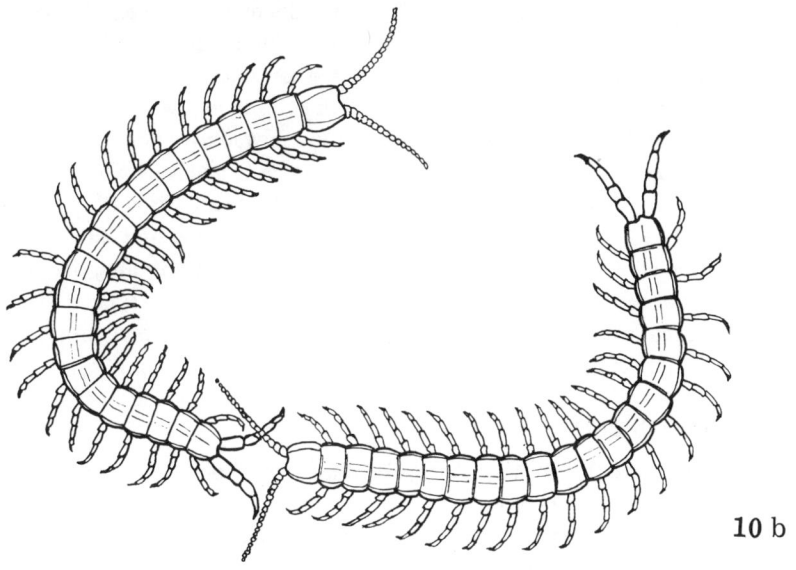

10 b

Figure 10. a. Two adult combative males of *Cormocephalus anceps* grappling with each other.  b. a female after courtship with the male follows him, tapping and stroking his end-legs with her antennae (after photos by B.Brunhuber).

Although the activities of centipedes can be watched under red light, their nocturnal habits and sensitivity make such observations very difficult and Dr Brunhuber in her thesis says rather ruefully that 'it took many months of continued observations combined with a bit of luck' before she was able to piece together a satisfactory account of the mating habits of *Cormocephalus anceps.*

35

In her laboratory she placed a male and female in separate boxes which were however joined by a short plastic tube 8 cm in length, the 'tunnel'. After a meeting of the partners sex intercourse began by mutual tapping with their antennae of each others end-legs (Figure 10); after about four hours the male moved away to the tunnel followed by the female, still tapping; when another hour had passed in this way the male protruded his genitalia, lifting the posterior end of the body, swaying up and down and sideways, touching the sides of the tunnel with the conical 'spinner', a specialised part of the genitalia used for this purpose, emitting a secretion which hardens rapidly in the air, producing delicate sticky threads like those of spider silk. In about 20 minutes a fairly thick wad of silk was formed and he then brought the spinner sharply down on the web, depositing a soft white spermatophore on it. The male then moved away and groomed his genital region while the female moved forward over the web tapping it with her antennae until she touched the spermatophore. She then moved quickly over the web with her genitalia protruded and picked up the spermatophore but the precise manner in which this was done unfortunately evaded the observer; after leaving the web the two sexes on meeting again resumed their customary aggressive attitude.

In another pair a somewhat different pattern of courtship was observed. The male grasped the female with his end-legs and moved over her, holding her along the length of her body while stroking her antennae with the tips of his end-legs; if he stopped the female rapidly tapped his end-legs with her antennae and he would resume the stroking (Figure 10); some time later the male constructed a web several times over but the female, not finding that a spermatophore had been deposited, destroyed them all and after some hours the discouraged male abandoned the web without depositing a spermatophore.

## 6. The transfer of sperm to the female

The spermatophore (Figure 12) is a small kidney-shaped object, 1,5 mm long and about half as wide; its tough covering envelope consists of three layers staining different colours and when expelled by the male is covered with a coating of sticky mucus which according to Klingel enables it to adhere transversely across the female genital opening (Figure 11). It is not known with certainty how the extremely large spermatozoa with their very long thread-like tails are enabled to emerge from the envelope; it may be that when a spermataphore is fixed in position, the pressure set up by the contracting envelope due to the drying of the mucus coating is enough to expel the sperms at a slow rate. Dr Brunhuber was the first zoologist to make a discovery which provides a more objective explanation. She found that the thickened convex border of the spermatophore was riddled with innumerable minute pore canals while the opposite concave margin was much less resistant, being provided with two circular thin-skinned areas in the middle. When the spermata-

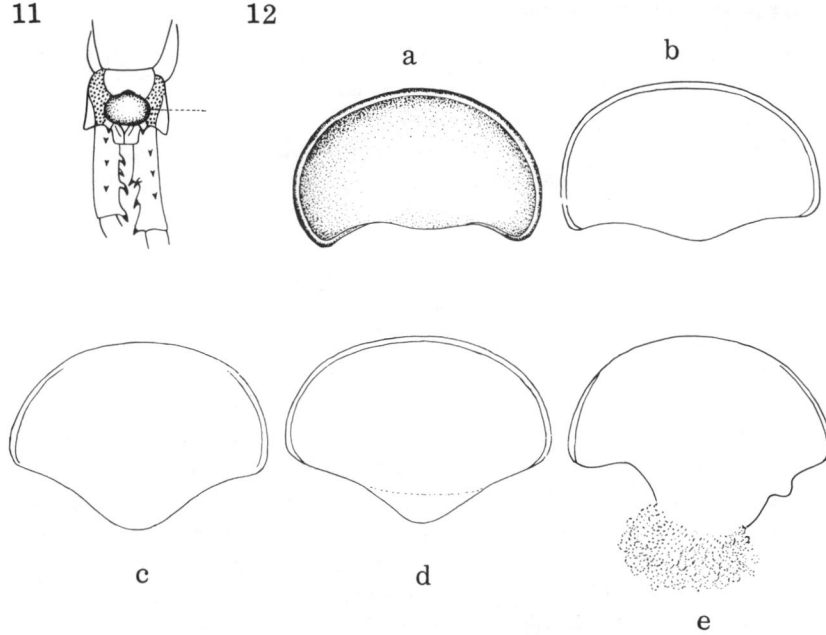

a

b

c

d

e

Figure 11. A spermatophore of the scolopendromorph *Otostigmus* placed sideways against the genital opening of the female (after J.M.Demange).

Figure 12. A freshly expelled spermatophore of *Cormocephalus anceps* (a) when placed in distilled water begins to swell at a weakened part of the envelope, the 'blister'; stages 2-4 show a progressive swelling until after two hours (e), the blister bursts, releasing the sperms (after photos by B.Brunhunber).

phore was placed in distilled or tap water these areas of less resistance commenced to swell and after 30 minutes had formed a small transparent blister which gradually grew in size until after two hours the swelling burst and the sperms slowly oozed out (Figure 12). Thus a hypotonic solution entering the spermatophore through the pore canals could rupture the weakened areas of the envelope by osmotic pressure. How such liquid would be obtained in the natural environment is an unsolved question unless the female at this time deliberately seeks water and by walking over it conveys enough to the spermatophore to bring it to the bursting point. It is rather doubtful if secretions of the female accessory glands, as suggested by L.Fabre in 1855, or other structures associated with the genitalia, could provide sufficient liquid to cause the release of the sperm.

37

## 7. Egg-laying and maternal care

Like all centipedes the Scolopendromorphs lay eggs and in comparison with many other lower creatures the number is small; on the other hand the eggs themselves are usually large and oval in shape.

Mating probably takes place in late summer. Egg-laying in *Cormocephalus anceps* occurs throughout August and September, that of the Natal forest centipede, *Cormocephalus multispinus* in October and November. Dr Brunhuber found that the number of eggs laid in a batch was on an average 30, while in the Natal forest centipede it was 23; such discrepancies can probably be attributed to the fact of the one species living in an area of winter, the other in an area of summer rainfall.

The Natal forest centipede chooses a damp, dark, and well-protected situation, usually under the side of a decaying log which has been softened by attacks of fungus; here a roughly hollowed out chamber is made, for the most part by her own bodily movements, or she may use an old nest or burrow vacated by some other small animal. When the dull white, translucent eggs appear they are coated with a sticky liquid from one of the sex glands, enabling them to stick together in a cluster round which the mother coils herself, protecting the eggs with her legs so that they appear to be enclosed in a little basket (Figure 13). The secretion which coats the eggs has to be renewed from time to time in order that they may be kept continually moist; it would be fatal for them if they were allowed to become dried out. The eggs are never permitted to touch the ground; small crumbs of soil would cling to their sticky surfaces and infect them with harmful bacteria and fungi, a mishap which often occurs when trying to hatch out batches of eggs under the artificial conditions of the laboratory; such secretions thus probably play the part of a germicide and disinfectant. More than one attempt has been made to rear young centipedes from the egg without the expert aid of the mother; all have failed as it has not been able to devise a substitute for the secretions from the glands of the centipede which appear to keep the eggs at the right degree of humidity and temperature; they either dry out and shrivel up or are attacked by the deadly moulds and fungi which destroy so many small animals living in damp and decaying wood.

The centipedes of this group show a remarkable solicitude for their young, rivalling the mother-care of many spiders and scorpions. The protective instincts of the mother must be very strong for she never leaves her brood during the six to eight weeks that follow the laying of the eggs, neither is she able to capture other animals for food in such cramped quarters. Previous to the preparation of the brood chamber, however, the female has been feeding generously and has stored a large amount of fat in her body in preparation for the long period of abstinence which is to follow. A German naturalist tried the experiment of substituting a pebble of the same size and weight for the egg mass; the mother accepted and guarded this with her body for five days, but after that, having presumably discovered the deception, she disappeared and was seen no more.

If the mother is disturbed in her nursery she tries to defend her brood by brandishing her spiny end-legs, or if the distractions continue, by making attempts to bite at the intruder; after certain limits to her patience have been reached she gives up the struggle and makes off, leaving the eggs to their fate. Sometimes a curious thing happens if the teasing is continued too long; perhaps as a result of the thwarting of her natural instinct the mother turns on her own eggs or helpless embryos and eats them one by one. When trying to photograph a mother centipede and her brood of motionless embryos in the forests of the Drakensberg, the writer took about 15 minutes to adjust and focus the camera, and though from time to time he noticed that the head of the mother centipede was moving in a strange rhythmic manner, he did not suspect that anything was amiss until the time came to click the shutter, when he found that of the original 19 embryos only two remained; the rest had been steadily disappearing down the mother's gullet.

The development of the centipede from the egg is marked by three stages which have been called the embryo, the foetus, and the adolescent stages. In the embryo stage the young are grub-like and motionless, and their bodies are still covered with a coating of liquid; in the adolescent stage they can be clearly recognised as centipedes and are quite active, while their skins are dry.

The embryo breaks open the almost transparent egg membrane, probably with the aid of a cutting tooth or egg-tooth, a small tool well suited for this purpose; there are two of these minute black teeth, one on each side and close behind the head, provided with a sharp chitinous edge. While still enclosed in the egg membrane there has already been considerable development; the tiny bud-like legs can be clearly seen and the antennae are composed of a number of as yet ill-formed segments. The rudimentary legs are the same in number as the adult centipede.

After emergence from the egg the embryo at once casts its skin, which also carries away the now superfluous egg-tooth; at this stage it appears as a small, white, worm-like object, bent upon itself in the shape of a horse-shoe, the head almost meeting the tail (Figure 13). It is still very delicate and soft, quite incapable of movement and dependent for food upon the yolk which has been brought with it from the egg stage in its immature stomach. It now casts its skin again and transforms into the second or foetus stage; here the little centipede is stretched out straight, and though still quite colourless and very soft, it is able to make a few simple movements; the legs now have tiny claws. It sheds its skin for the third time and enters upon the last or adolescent stage which is reached after about 1½ months; in this, as implied by the name, the main structures of the body, such as the mouth parts and claws, are present in their final condition, though as yet on a small scale and still soft and unchi-

39

tinised. The young centipede begins to creep restlessly about, entwining itself with its brothers and sisters but keeping close within the embrace of the mother, who all this time has been guarding the developing brood with the shield of her body (Figure 14).

From now onwards the young tend more and more to manage for themselves but they moult several times more before they finally leave the mother's shelter; after each moult they are a little darker in colour; the openings of the breathing organs, the spiracles, begin to appear and the poison glands become ready to fulfill their function. They have carried over from the egg and embryo stages a certain amount of food in their primitive stomachs in the form of yolk which can be easily seen through the semi-transparent cuticle. As this diminishes they begin to drift away in ones and twos, the urge of hunger driving them in various directions to seek and capture their own food, never returning to the nursery or meeting again, except as an enemy, the parent who has watched over them with such unceasing care. After the 10th or 11th moult and about six months after hatching, the colouring of the adult has been attained; they are sexually mature and full grown centipedes ready to bring a new generation into the world and repeat the cycle of living.

The life of the centipede from the hatching of the egg until its death has been estimated to extend over six years or even longer, a long span of time for such a small and lowly animal. At intervals it sheds its skin, usually in the summer and autumn months but unlike the millipede it does not undergo a period of difficulty and crisis at such times. The process also does not take so long, in adults only 20 minutes to an hour, as the skin at the back and sides of the head merely splits apart and the flap thus formed is turned forward, allowing the centipede to struggle gradually through the gap. Soon it wriggles free of the old and tattered skin, which it leaves behind with the end segments wrinkled into the front ones like the bellows of a concertina. The moulted skin is then eaten.

## C. THE LITHOBIOMORPHA OR STONE-CENTIPEDES
Type: *Lithobius* Stuxberg

### 1. *General description*

If length of leg were to be used as a criterion for distinguishing the four orders of centipedes, a rough and ready method would be to say that in the Geophilomorpha the legs are shorter than the width of the body, in the Scolopendromorpha equal or a little longer, in the Lithobiomorpha distinctly longer, in the Scutigeromorpha several times as long (Figure 17). In general the Lithobiomorpha and Scutigeromorpha, which differ from the other two suborders in their development as well as their structure, can be called long-legged centipedes as contrasted with the short-legged Geophilomorpha and Scolopendro-morpha. Together they are classed by systematists as a subdivision of the Chilopoda with the name Anamorpha while the Geophilomorph and Scolo-pendromorph centipedes constitute the Epimorpha.

The Greek name *Lithobius* or stone-dweller has been given to this type of centipede on account of the frequency with which it is found under stones but in South Africa it lives just as often under rotting wood or in forest leaf litter. It nearly always has a shiny appearance as if it had just been dipped into water and is light to dark reddish-brown in colour with hardly any pattern markings; younger stages of growth and specimens which have recently moulted can be distinguished by their paler colour. (Fig.16, p.42)

The shape of this centipede is quite distinctive and its body is much shorter than either of the two groups which have already been described, with only 15 dorsal shields or tergites; these instead of being uniform in size as in the Geophilomorpha or nearly so, as in the Scolopendrids, differ markedly, a larger shield usually alternating with a smaller. This arrangement of the dorsal shields is one means of distinguishing stone centipedes from the other orders. The antennae at the front of the head are long and slender, with anything from 20 to 50 joints thickly covered with sensitive tactile hairs, and in some cases also with an organ of smell. There is usually only a single eye in the South African stone centipedes though in one species there are none, in another a cluster of about 12 ocelli. The 15 pairs of legs are distinctly longer than those of the large Scolopendra centipedes and towards the hind end of the body they begin to increase in length, the last leg being much longer than any of the others. On the underside of the four last pairs there are rows of small round pores which are the openings of glands secreting a sticky liquid used as a means of defence (Figure 16).

## 2. *Devices for defence and escape*

The stone-centipedes are energetic, quick-moving animals, darting here and there, and when found hiding under a stone they usually run round again to its dark underside. When cornered they sometimes sham dead and lie still with the body curled up. The last pair of legs is not used for running but is lifted up in the air or trailed along the ground like the steering oar of a boat; the legs are very easily discarded and one is lucky to be able to catch an unin-jured specimen with the full number of legs. The loss of these legs (autotomy) is not accidental but provides a means of escape should the owner be seized by one of its enemies; in the same way some lizards, like the geckos, are able to dispense with their fragile tails in an emergency. The break occurs at a spe-cially weakened joint near the base of each leg, and a ring of sphincter muscles which closes round the wound and prevents too great a loss of blood. The lost leg is soon regrown and at the next casting of the skin a new one appears to take its place.

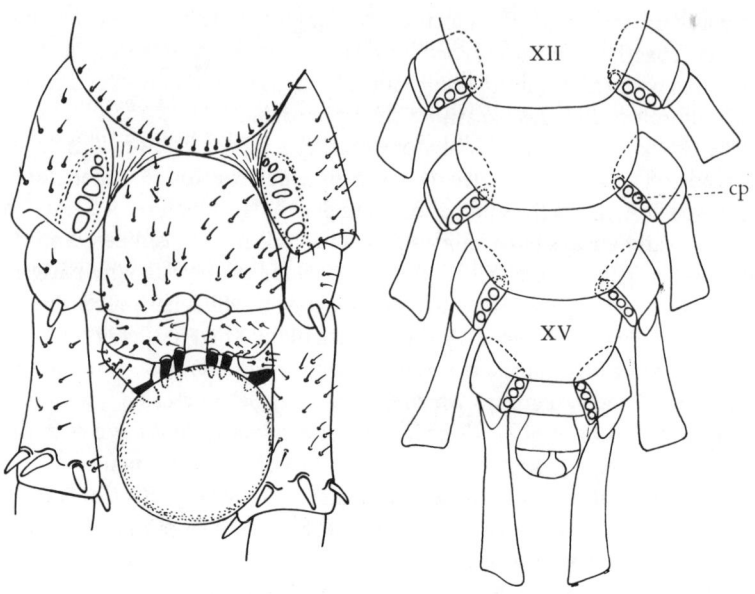

Figure 16. Left, the tail end of a *Lithobius* female holding the large egg with the claw-like clasping appendage. Right, rows of coxal pores on the four last legs of *Paralamyctes.*

The stone centipedes practise quite a number of ways of outwitting their enemies. In the first place their dull brown colouring, which is very like their earthy background, makes it difficult for them to be seen. When circumstances are favourable however they rely on their speed and agility to make good their escape; they seldom run backwards like the earth centipedes, only resorting to this method of retreat when they are enclosed in a narrow passage under a stone, or in cracks of the ground where they are unable to make a direct turn; when cornered by an enemy in such a restricted space they cannot escape by flight, even at the cost of sacrificing a leg or two. There is however still another weapon in their armoury; at the base of the last four pairs of legs can be seen one or two rows of small round pores looking like the holes of a sieve (Figure 16); when the centipede is approached by another creature with aggressive intentions it turns its back on him and lifting the back legs high in the air squirts a jet of sticky liquid from these pores which in many cases either drives him off or successfully entangles him in sticky threads. A German naturalist, Dr Karl Verhoeff, noticed that a pugnacious species of ant, which tried to attack a *Lithobius* and was treated in this fashion, took more than an hour to release itself from the tangling sticky threads. The chief enemies and competitors of these centipedes in their search for food are the small wandering creatures which live on the surface of the

42

ground, such as wolf-spiders, ants, and ear-wigs; a stone centipede was observed by Dr Verhoeff to overcome wolf-spiders (Lycosids) on three occasions by its speed and agility alone; the contests all ended by the centipede biting the spider in the stomach after which it proceeded to devour the soft parts of the victim. As a rule however *Lithobius* feeds on small flies and gnats, avoiding insects such as beetles which have a hard carapace; only the soft parts of the flies are eaten, the hard parts such as the thorax being scooped out and then discarded. Stone centipedes seem to be thirsty animals and water is evidently necessary to their existence as they have often been seen to lick up drops of water from the ground. After feeding, the antennae, body and legs are carefully cleaned in the manner that has been described in the case of *Scolopendra.*

## 3. *The poison*

The victims of the stone centipede are seized and pierced with the large poison claws so that they are immediately killed or paralysed by the poison, which seems to be very similar in its effects to that of the spider. As far as human beings are concerned the bite of the creature is not a cause for anxiety, seeing that the poison claws are in most cases too small and weak to pierce the skin. Only one large species, found in the islands of the Mediterranean, has been known to bite human beings and such bites were not followed by any serious consequences.

## 4. *Courtship and mating*

The sexes in *Lithobius,* according to the French authority, Demange, pair off in early spring; there is more association between the sexes than in the case of the Scolopendromorpha but also more aggression between members of the same species. In encounters between the sexes the males acquit themselves better than their fellows of the Scolopendromorpha and it is not the male that usually falls a victim to the female; the results of such contests may be described as fairly even.

In a sexual encounter the male indicates his readiness to lay down a spermatophore by bobbing his posterior up and down against the female. She on her part shows her willingness to take up the spermatophore by following the male and remaining in touch with him as he leads her in a kind of love promenade (Liebesspaziergang). After spinning a web and depositing a viscous matrix upon it the male pushes a drop of sperm into it and then moves back and spins several sticky threads, all pointing in the same direction, the 'Signalfaden', which serve as traffic signals or guide lines for the female. He then moves towards her, tapping her antennae with his anal legs, causing her to move forward over his posterior end so that she lies in front of him with the genital

43

segment protruding. He flattens himself so that the spermatophore becomes visible between his parted anal legs and the femal then picks it up with her gonopods, carrying it around for some time, during which presumably the spermatozoa are absorbed into her genital orifice.

## 5. *Egg-laying and development of the young*

The way in which the young are brought forth and their development is rather different from that of the Scolopendromorph and Geophilomorph centipedes. Eggs are laid but there is a total absence of the maternal devotion to the offspring which is such a remarkable feature of the other two groups. Instead of a large batch of eggs, these arrive one at a time with a considerable pause between the appearance of each egg; they are laid in the late spring and are small and few in number. When an egg appears it is carried around for some time before being coated with a sticky secretion; it is then grasped and held by the gonopods, a paired pincer-like organ or clasper situated near the genital opening, which is specially fitted for this purpose (Figure 16). Males kept in captivity with females have been said to show cannibalistic tendencies at this time for they make a rush at the female and try to snatch the egg from her and devour it. As soon as the egg is firmly gripped by the pincer-like claspers the female hurries off to some convenient spot where the egg is rolled round and round until completely covered with small particles of earth clinging to its sticky surface. When this operation is complete the egg resembles a small round ball of mud indistinguishable from the surrounding soil and safe from predators and the unnatural appetite of the male; it is then dropped to the ground and left to its fate. It may perhaps be considered doubtful if the male displays these infanticidal propensities in natural conditions where there would be more room to move about and plenty of the food to which these centipedes are accustomed.

The two long-legged orders, Lithobiomorpha and Scutigeromorpha, the Anamorpha, differ from the two shorter-legged orders previously described, the Epimorpha, in an important aspect of their development. While the larvae of the Epimorpha emerge from the egg with the same number of legs, fully formed but rudimentary, as the adult, the lithobiomorph and scutigeromorph embryos are born with a small fixed number and further pairs are added intermittently at each moult as the larva grows.

When the young stone-centipede breaks open and emerges from the egg it is in a very immature condition and unable to take food. There are only seven pairs of legs instead of the 15 pairs of the fully grown centipede and the embryo is able to make only a few simple movements. Four larval stages follow in each of which one or two new pairs of legs are added to the original number, so that the total of legs is progressively increased. At the end of these four stages the young centipede is still very far from being full-grown, or with all

44

its organs complete; at least four more stages, each accompanied by a moult, must be passed through before the young are mature and able to reproduce their kind, the whole development from egg to maturity taking no less than three years to complete. It is therefore not surprising that one nearly always finds under the same stone a number of centipedes of different sizes with darker or lighter colours, an indication that they are at various stages of their development. The same female may go on laying eggs for several years and throughout the five or six years of her life she sheds her skin at fairly regular intervals.

The stone-centipedes are for most of the time light avoiding stay-at-homes but at night they wander forth in search of their prey. For their well-being they need a certain amount of moisture in the air they breathe and this they find under stones or among the damp decaying leaves of forests. In our country they seem to be equally at home in both these habitats though in arid regions, such as the Kalahari and Karoo, they live almost entirely under stones and in greatly restricted numbers. The smallest of these centipedes in other parts of the world when full grown measures only a few millimetres, the largest about 5 cm; the biggest species in South Africa, 3-4 cm long, is the handsome reddish-brown *Paralamyctes spenceri* which is found in all the coastal and montane forest of the Republic, from sea-level to about 3 000 m of altitude.

## D. THE SCUTIGEROMORPHA OR HOUSE-CENTIPEDES
Type: *Scutigera* Lamarck

### 1. *General description* (Figures 17 and 18)

The Scutigeromorpha, though related to the previous order, in many respects stand apart from all the other three in their organisation and way of life. *Scutigera,* the type animal, is the most advanced member of the Chilopoda; it has the most acute eye-sight, the best developed mouthparts (mandibles) and is by far the most speedy runner of all the centipedes.

These chilopods derive the name *Scutigera* or shield-bearer from the eight rounded tergites or shields on the back of the body but they might also be called house-centipedes as they come into houses more frequently than do any of the other kinds. They are moderately small centipedes but have 15 pairs of immensely long legs by which they seem to be surrounded as by a palisade, and these legs give them the appearance of being larger than they actually are (Figure 17). The antennae at the sides of the head are also very long, delicate, and whiplike, being composed of a very large number of small joints. The large eyes differ from those of other centipedes and resemble the eyes of insects in being not single ocelli but compound or faceted eyes (Figure 18a). The surface of the eye is composed of about 100 tiny six-sided facets,

45

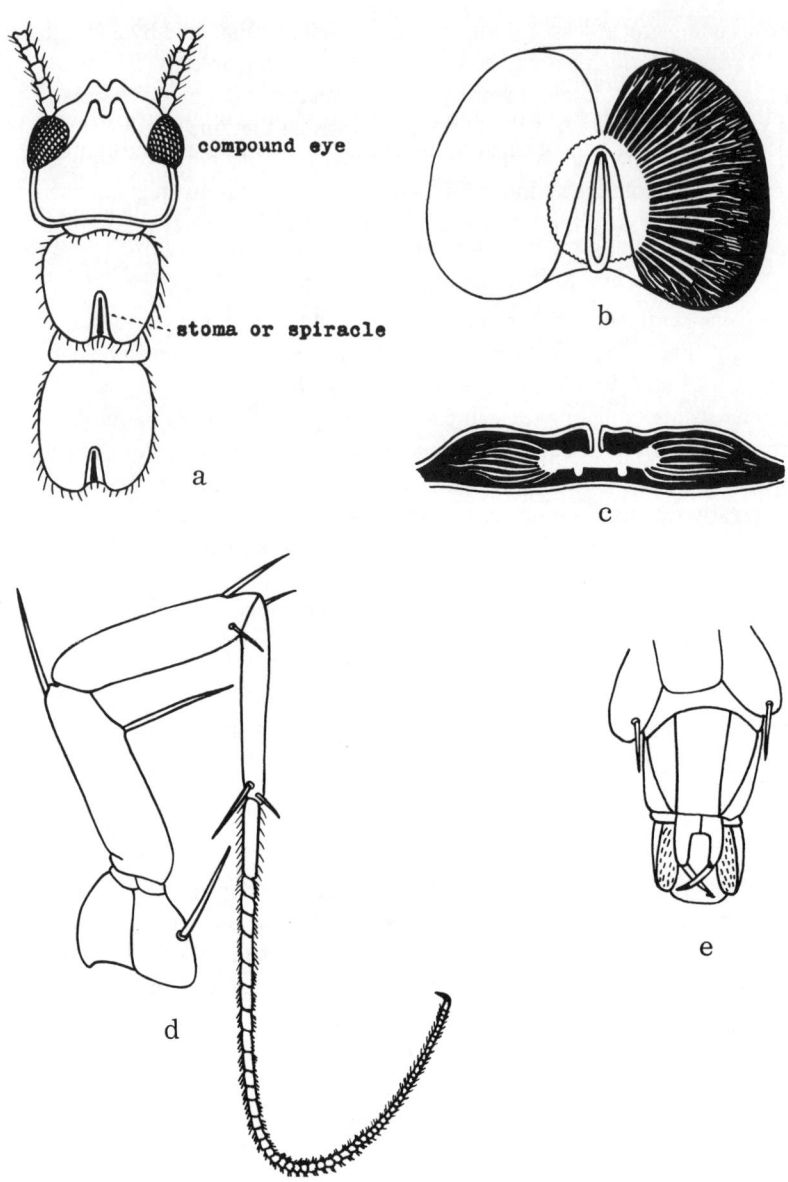

compound eye

stoma or spiracle

a

b

c

d

e

Figure 18. *Scutigera coleoptrata*. a, the head and two body segments; b, one of the stomata with the right half of the dorsal shield removed to show the tracheae; c, cross section of a segment to show the tracheal system; d, one of the legs; e, posterior end of a female with the pincer-like claspers for holding the egg.

each provided with a lens and a retina; there can be little doubt that eyesight is much better developed in this group than in any of the other centipedes and that *Scutigera* actually uses its eyes to see moving objects. This is evident from the way in which it craftily approaches and leaps upon its prey which consists for the most part of small flying insects such as flies and mosquitos. The legs and the tarsi especially, are very long and flexible like the lash of a whip and resemble the antennae in appearance; the tarsi are again divided into a large number of minute joints, sometimes as many as 400, each with its own muscle attachment, giving them such mobility that they can be used like the lassoos of cowboys to fling around the prey. (Fig.18d).

The manner of walking and running of these centipedes is very unusual in the arthropoda; it is plantigrade, or walking on the soles of the feet rather than with only the ends of the tarsi touching the ground (digitigrade). Such a mode of walking is also employed by some of the harvest-spiders (Opiliones), the long-legged order of Palpatores; in these arachnids the extremely long and flexible tarsi may have as many as 80 segments and are used like the prehensile tails of monkeys for curving round the grass stems which they grasp as they run with great speed and sureness of foot.

*Scutigera* loses its legs even more easily than the stone-centipedes do theirs; the last pair seems to be especially fragile, and at the lightest touch they break off at a specially weakened joint near the base of the limb; when this happens the legs still have some power to keep on moving jerkily for a while, just as the broken off tails of geckos keep on twisting and wriggling as if they were alive. In a short time a new leg is grown to take the place of the missing one.

The mouth-parts are strong and of an advanced type with extremely large mandibles especially well fitted for dealing with the hard cuticles of insect prey; the first pair of maxillae have a peculiar brush of setae attached to them which can be pushed outwards for cleaning appendages such as the legs and antennae; the second pair of maxillae are much stronger, heavily spined and more leg-like than they are in the other orders. There is an exceptionally large pair of poison claws or toxicognaths.

## 2. *The unique breathing organs of Scutigera*

The respiratory system is the most peculiar of any found in the Chilopoda or for that matter in any air-breathing arthropod(Figures 18b,c). Instead of the usual pairs of tracheal tubes at the sides of the body for bringing air direct to the organs, this centipede has single openings in the middle of the back, one on each of the tergites except the last. They are large slit-like apertures resembling on a much larger scale the respiratory pores or stomata of plants, and have been called by the same name. That there is a flow of air into the stomata can be demonstrated by placing a droplet of water near one of the openings; in less than a minute it will be absorbed and disappear. Inside each of the stomata there are numerous short tube-like tracheae branching out from a central chamber so that oxygen passing through their walls is diffused into the blood stream (Figure 18b). The system probably resembles that of terrestrial crustaceans such as woodlice more than any other type of breathing organ; in the woodlice (Isopoda) the tracheal trees or pseudotracheae consist of tufts of minute trachea-like tubes leading into a common chamber with a slit-like opening on the abdominal appendages attached to the under surface of the animal. There is thus a considerable difference between the system of *Scutigera* and that of insects, other centipedes, millipedes and so on.

In *Scutigera* the blood plays a more important part, carrying the oxygen which it has received from the tracheae to the various organs as in the higher animals, a method which is an improvement upon that of other centipedes and millipedes where air is carried to the organs by means of a rather poorly developed tracheal system and the blood plays very little part in its distribution. It is not surprising therefore that the display of restless energy in *Scutigera* far surpasses that found among any of the other centipedes; taking its small size into account it is one of the fastest moving of all terrestrial invertebrates. This is also due of course to the very long legs which increase in size from in front backwards until the last pair reaches about twice the length of the whole body. The legs themselves, though slender, are strong, with mechanically efficient joints; instead of extending flatout sideways they are bent sharply in the middle, serving to support the body more easily and to lift it clear off the ground. The body is cradled or suspended between the legs as in the early horse carriages where the body of the vehicle was supported between the wheels by leather straps. The segments of the body are rigidly attached to each other, lacking the intermediary segments of the Geophilomorpha which promote flexibility; in its lightning dashes to and fro the fleet-footed *Scutigera* does not make undulatory movements in order to get round obstacles in its path, instead the body is lifted over them.

### 3. *The manner in which the prey is captured*

Scutigeromorph centipedes are nomadic hunters and live mainly on flies and mosquitos; they leap upon them with incredible swiftness using their long, flexible legs almost like lassoos; it is not unusual to see a *Scutigera* pounce on three or four flies at one swoop, surrounding them with a palisade of legs within which the victims are enclosed like hens in a coop; while one fly is being eaten, others are held down with the legs. The captured insects are quickly pierced with the poison claws but only the soft parts are eaten, the hard skeleton of chitin being rejected.

48

## 4. *The situations in which they are found*

Although they are often to be seen in daylight the greater part of their activities take place at night. They often come into houses at night or in the early morning and domestic utensils containing water seem to have a great attraction for them as they are often found in pails and baths. The writer used to find one in his bath at Sea Point, Cape Town, on most mornings in the summer time; it may be that they came there to hunt for flies; it is also not improbable that they needed the water to drink. One naturalist says that according to his observations they sometimes come into houses in order to hunt for bedbugs living in the cracks of the walls. In Australia, where they are often found in houses, the common name for them is 'bugkillers', and they no doubt play quite a useful part as destroyers of such pests as flies, mosquitos, and silver-fish. Under natural conditions they retire under fallen logs and pieces of rotting wood, and in Natal they have been found under the decaying trunks of *Aloe ferox* which have fallen to the ground. The house centipedes are almost entirely inhabitants of the warm tropical and subtropical lands. There are very few species and the most well-known of these, *Scutigera coleoptrata,* is found in many parts of the world as well as in South Africa, more often in the neighbourhood of sea-ports as is usual in the case of imported animals. It is our commonest house centipede and was probably introduced at some time into this country.

## 5. *Grooming habits*

The Scutigeras, like the Lithobiid and Scolopendrid centipedes, clean themselves very conscientiously and thoroughly with the aid of the brushes on their mouth-parts. They have many skin glands, especially on the legs, which exude an oily substance; this is spread in a thin coating over the cuticle and keeps it free from dirt and harmful microscopic organisms.

## 6. *Courtship and mating*

*Scutigera* illustrates the final and most advanced stage of the association of the sexes in centipedes with regard to the courting and mating rituals. The male no longer lays down webbing as a guide to the female but takes over this rôle himself. As in other centipedes there is at first the mutual tapping by the partners with the antennae but after this the courtship is considerably more elaborate, with the sense of sight coming into play.

The male by suddenly pushing himself under the body of the female reveals himself to be sexually prepared while the female responds by raising her body off the ground. This play, repeated many times, leads on to another with the male bobbing up and down in front of his *vis-à-vis.* When the female shows interest the male deposits a spermatophore on the ground consisting of a matrix in which the mass of sperm, unprotected by an enclosing envelope, is implanted. Once the spermatophore has been deposited the male pushes the female over it whereupon she takes up the mass of sperm and inserts it in her genital opening with the aid of the end-legs; in some cases the male may do this himself but, as far as is known, there is no copulation. The process recalls in some of its aspects, the manner of sperm transfer in the scorpion.

### 7. Egg-laying and development of the young

The birth and development of the young are very similar to what has been described in the case of the Lithobiomorpha. The eggs are laid singly and held by the genital appendages, the gonopods, specialised structures in the female resembling a small forceps (Figure 18e); as they appear they are covered first with a sticky secretion and then with small crumbs of soil, after which they are dropped casually to the ground. The larva emerges from the egg in a very helpless and incomplete state with four pairs of crude and rudimentary legs; it is then only a tenth of an inch (2.5 mm) long but goes through nine more stages, increasing gradually in size and adding more legs at each shedding of the skin. In the first three stages it is still absorbing the yolk which has remained with it from the egg, but from the fourth stage onwards it is able to take other food from the outside. After a year the larva has reached its full stature and becomes a mature centipede.

*Scutigera* resembles the insect in its habits of life and its structure more than do any of the other centipedes and is certainly one of the most curious members of the vast and diverse army of jointed animals.

### 8. The four progressive stages of the Chilopoda (Figure 19)

The four progressive stages from a vermiform (*Geophilus*) to an insectiform centipede (*Scutigera*), illustrated in Figure 19, are the outward expression of drastic differences in the habits and manner of life of the four orders. The Geophilomorpha are examples of almost totally subterranean animals, leading a burrowing existence, seeking their prey underground. The Scolopendromorphs live largely on the ground surface but in crevices and holes or under stones where they can follow the prey. The Lithobiomorpha are actively predacious but nocturnal and partly cryptic inhabitants of such situations as forest litter, the undersides of stones and decaying tree trunks. The fleet-footed *Scutigera* stands at the opposite extreme of *Geophilus,* being a chilopod which lives perpetually above ground, a coursing predacious centipede, able to pounce or change direction quickly. It is not cryptic or particularly light avoiding and with its excellent eye-sight can pursue the prey in daylight.

In general these stages represent changes from centipedes with long slender bodies composed of numerous segments to those with short bodies and few

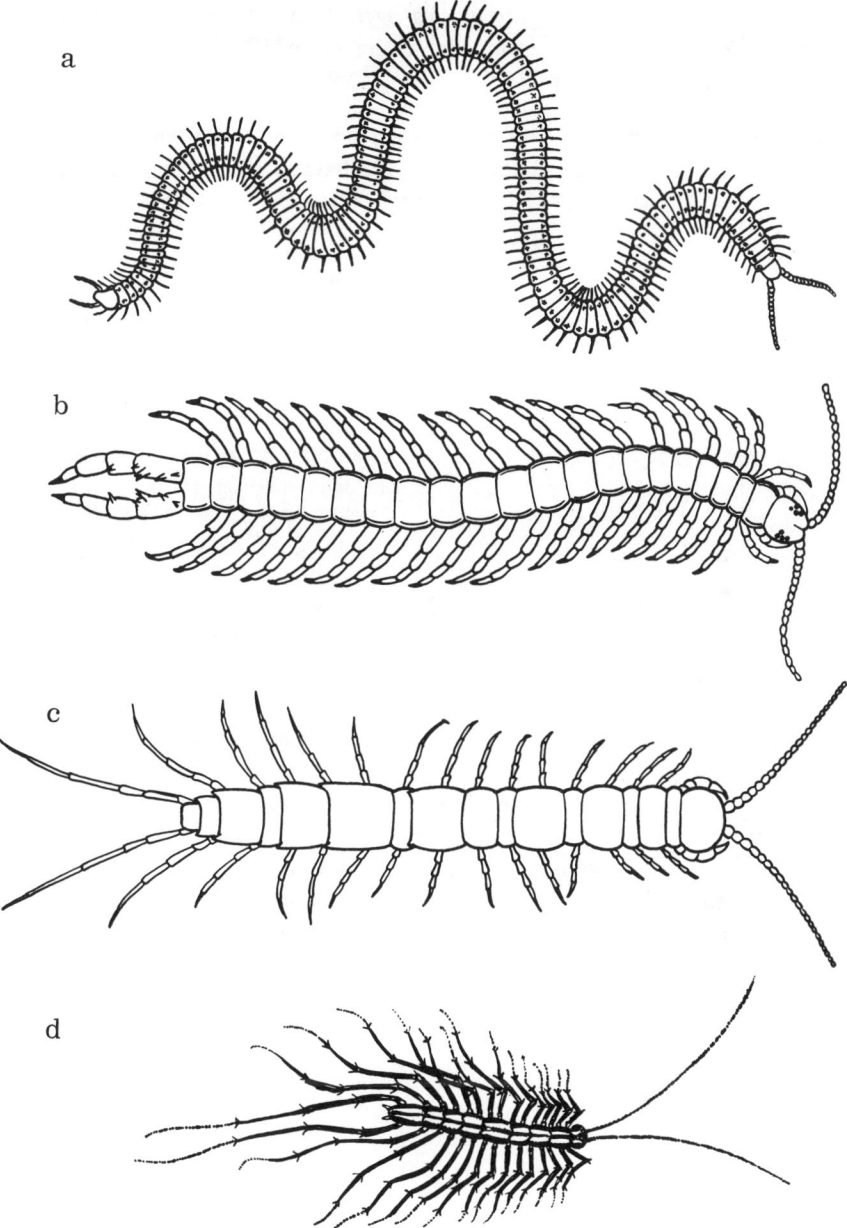

Figure 19. Types of the four orders of centipedes: a, earth-centipede, *Geophilus* (Geophilomorpha); b, large-centipede, *Scolopendra* (Scolopendromorpha); c, stone-centipede, *Lithobius* (Lithobiomorpha), and d, house-centipede, *Scutigera* (Scutigeromorpha).

51

segments, and from short-legged to long-legged forms. Some of the other structural differences of a less fundamental nature, but which also constitute a progressive series, are shown in the following table.

| Character | Geophilo-morpha | Scolopendro-morpha | Lithobio-morpha | Scutigero-morpha |
|---|---|---|---|---|
| Spination of legs | Legs without spines | Legs with very few spines | Legs with fairly numerous short spines | Legs with many large, long spines |
| Number of tarsal segments | Legs with one tarsal segment | Legs with two tarsal segments | Legs with one to three tarsal segments | Legs with numerous tarsal segments (up to 50) |
| Number of antennal segments | 14 | 17-20 | 19 to over 100 | With about 400 |
| Eyes | Absent | 4 simple ocelli or none | A cluster of 14 simple ocelli, 1 or 0 | Two very large facetted (compound) eyes |

The same changes take place in lizards; lizards which have completely lost their legs, depend very largely on undulations and twists for moving rapidly over the ground while those with weak shorter legs and slender elongate bodies also progress with serpentine movements. Lizards on the other hand with normal, moderately long legs, have shorter bodies and move in a fairly straight line.

In a cramped living space, in narrow crevices or under the ground itself, legs are more of a hindrance than a help; animals which live in the open generally have to move faster and long legs are more useful to them. The same changes in the length of the legs and movements of the body take place in two wholly different creatures, centipedes and lizards.

Figure 2. The common and widespread centipede, *Cormocephalus nitidus*, with the end-legs held in defensive attitude (photo R.A.Holliday).

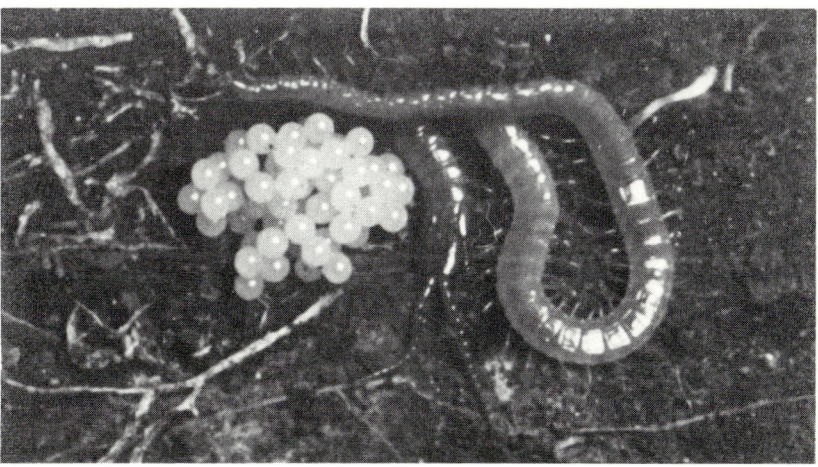

Figure 9a. The female geophilomorph centipede *Brachygonarea*, with her cluster of eggs (photo R.A.Holliday).

a

b

Figure 13. Above, the centipede *Cormocephalus multispinus* curled round her eggs in the brood chamber. Below, an early stage in the development of the young, the immobile comma-shaped embryos.

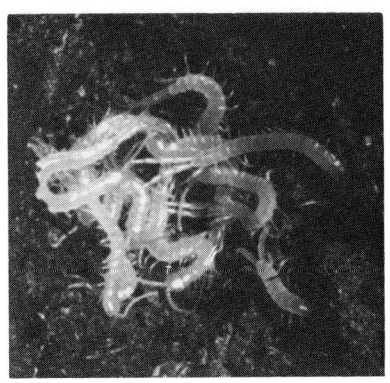

Figure 14. Left, a further or adolescent stage; the bodies of the embryos are now straight though still soft and colourless. Right, after another moult they are larger and more active, ready to leave the mother (photo R.A.Holliday).

15

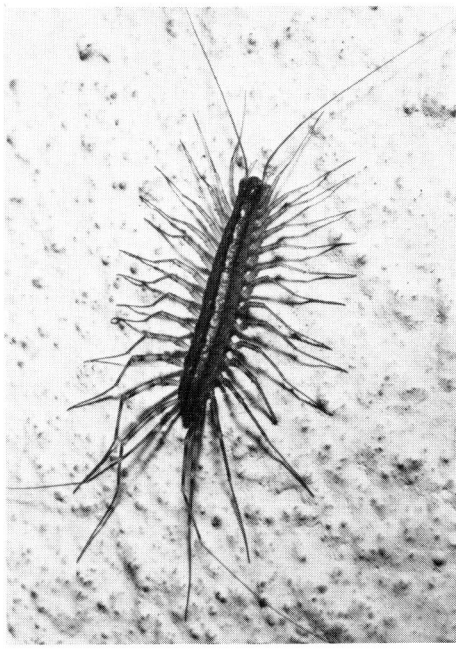

17

Figure 15. *Paralamyctes spenceri,* the commonest and largest of the South African Stone-centipedes (Lithobiomorpha) (photo R.A.Holliday).

Figure 17. The house-centipede, *Scutigera coleoptrata* (photo D.D'Ewes).

a

b

Figure 22. Above, a millipede walking and below, in defensive spiral enrollment (photos R.A.Holliday).

a

b

Figure 24. Above, the nuptial embrace of the millipede *Doratogonus,* below, the moulting of the skin in *Chersastus* (photos R.A.Holliday).

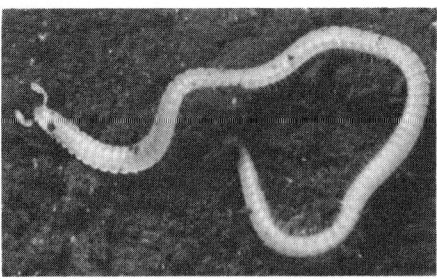

Figure 26. The widespread spirostreptid millipede, *Doratogonus flavifilis* (photo Roy Smith).

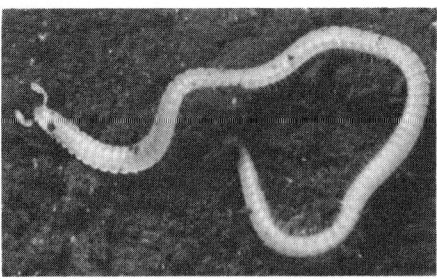

Figure 36a. The Natal sucking millipede, *Nematozonium* (Colobognatha)

Figure 40a. The South African 'garden centipede', *Hanseniella capensis*

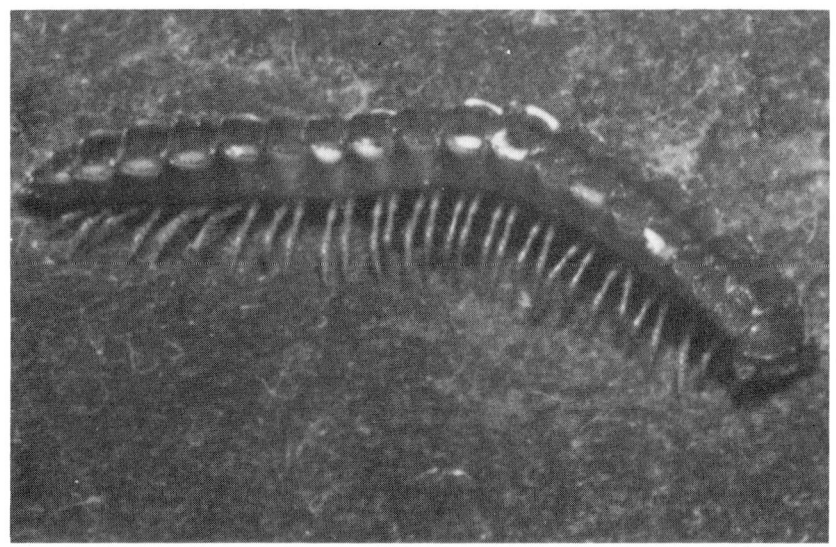

a

b

Figure 30. Two keeled millipedes (Polydesmoidea). Above, *Platytarrus* with droplets of cream coloured secretion from the defensive glands on its back; below, *Gnomeskelus* in copulation (photos R.A.Holliday).

Figure 33. The giant South African pill-millipede, *Sphaerotherium giganteum* from Natal, walking, inverted and enrolled (photo R.A.Holliday).

# IV. CENTIPEDES OF THE PAST

Few fossil centipedes of any great age have been found and thus little light can be shed upon their ancestral history. In this respect the position is very different from that of the millipedes, since we have a large number of well preserved remains of these animals which lived in the forests of the Devonian and Carboniferous periods. This is to a certain extent understandable as the strong tests of millipedes reinforced as they are with calcium, lend themselves to preservation far better than do the softer bodies of centipedes.

As a whole the fossil centipedes of the older palaeozoic era are a very imperfectly preserved collection of remains. They have nearly all been found in the upper Carboniferous strata of North America and were described by Dr S.H.Scudder. There is now considerable doubt as to whether they are centipedes at all and a number of specialists on myriapods have taken the view that they may be marine worms with a large number of segments which in many cases have leg-like organs, or parapodia, for walking and swimming; some of these have a superficial resemblance to centipedes especially to the worm-like geophilids with their numerous body segments.

Be that as it may, among the fossil centipedes which have been described by Dr Scudder, one form, *Ilyodes* appears to resemble the present day *Geophilus*, another *Eileticus*, is more like one of the Scolopendrids, and a myriapod with few but elongated legs, *Latzelia*, certainly has the outward appearance of a centipede like *Scutigera*. All these however are represented by fragmentary petrifacts in which few of the leg or head structures have been clearly preserved and we must wait for discoveries of better material before anything of value can be said about fossil centipedes as a whole.

Coming back to more recent geological times, representatives of all the four centipede groups have been found in the fossilised amber of the Baltic which dates from the Oligocene (Tertiary) period, about 25 million years ago. These groups were represented by the genera *Scolopendra, Geophilus, Lithobius* and *Scutigera,* which differed very little, if at all, from the present day centipedes of the same genera.

Only one fossil centipede has hitherto been discovered in South Africa, in early Pleistocene deposits dating from about three million years ago. This first and only specimen was noticed in 1974 by a small girl of eight years, Laura Graves, on a visit with her parents to the Makapan caves site; she insisted that it was something unusual.

It was preserved in a small piece of calcite embedded in the grey breccia of the limeworks at the caves, famous for the large variety of fossil remains, both animal and humanoid, which have been found there. It has been studied by Dr J.W.Kitching at the Bernard Price Institute for Palaeontological Research, Johannesburg.

The specimen, though a juvenile only 30 mm in length, is well preserved, with some of the small structures such as the claws clearly discernible. It can without much hesitation be assigned to the genus *Scolopendra*, possibly *S. morsitans*, a species found throughout Africa and in most of the warmer regions of the world; the dark crossbands on the tergites which are characteristic of the typical *morsitans*, can be quite clearly seen.

# V. THE IDENTIFICATION OF CENTIPEDES

After some consideration, the writer has decided to give keys for the genera and not go into greater detail with keys to the species, which would almost certainly convert this part of the book into a treatise on classification. One genus of centipede is almost always quite distinct from another, even to some-one unfamiliar with this class of animals, while the species of any given genus are not so easily separated, the distinction being normally based on obscure and minute structures such as the absence or presence of spines, grooves, pores of the gland openings and such-like. The difficulties become greater when dealing with a large genus like *Cormocephalus,* with about 23 species.

In classifying any centipede specimen it must first be referred to one of the four large subdivisions of the Chilopoda which have already been separately described in the earlier part of this account. These groups are very different in appearance and can be easily distinguished by means of the illustration on page 31 and with the aid of the following key.

*A key to the four orders of Chilopoda*

1. Slender elongate and worm-like centipedes, with a large number of body segments and leg pairs, which may be anything from 39 to 128 in number but are usually far more than 39, the number variable even in the same species. No eyes. Antennae with 14 segments, the number constant. Movements slow and worm-like. Usually living a subterranean existence in soil but in semi-arid regions also under stones and the bark of trees . . . . . . . .　　*Geophilomorpha* (Earth centipedes)
2. Large centipedes with thicker, more robust bodies; number of legs never more than either 25 or 27 pairs, this number always constant in a given genus; eyes nearly always present, consisting of four small ocelli on each side of the head, absent only in the genus *Cryptops.* Antennae with 17-20 segments, usually 17. Movements moderately fast. Usually living under stones or in rotten wood, occasionally under the bark of trees . . . . . . . . . . .　　*Scolopendromorpha* (Large
   centipedes)
3. Small centipedes with only 15 pairs of legs, the number fixed in all species; the legs, especially the last two pairs, much longer than those of the Geophilomorpha or Scolopendromorpha. The 15 shields visible on the dorsal surface of the body unequal in size, forming a series in which a large plate usually alternates with a shorter one.

Eyes usually present, consisting of a group of small separate ocelli;
antennae with 13 to about 100 joints (in South African species
never more than 50) . . . . . . . . . . . . . . *Lithobiomorpha* (Stone-
centipedes)

4. Small centipedes with 15 pairs of legs, the number fixed in all
species; only 8 large shields visible on the dorsal surface of the body;
eyes large and consisting of a large number of facets like those of
insects; legs and antennae very long in proportion to the body, the
legs with numerous long and strong spines, the flexible tarsi composed of a very large number of joints; the antennae and last pair of
legs whip-like and longer than the whole length of the body;
antennae with very numerous (about 400) segments
. . . . . . . . . . . . . . . . . . . . . . . . . . . . *Scutigeromorpha* (House or
shield centipedes)

*The Geophilomorpha*

There are about 16 genera and 72 species of this group of centipedes in
Southern Africa but it is quite impossible to give a key even for the genera of
this group. The distinguishing characters depend upon dissection of minute
parts of the anatomy, usually the mouth-parts, and it would merely compli-
cate matters and confuse the reader to attempt such a key. The earth-centipedes have been very insufficiently collected and studied; the list of species
already known is probably only a small fraction of the total fauna which
will be found to inhabit South Africa. When the names of species are required
specimens would have to be sent to one of the few taxonomic specialists
living in various parts of the world.

The fauna of our region is composed almost entirely of small indigenous
inhabitants of the temperate rain forest; all the species have a restricted distribution and there are no introduced forms as in the other three orders; a
possible exception is *Ballophilus braunsi* which is widespread in South Africa
and is also found in Madagascar. *Ballophilus* can easily be distinguished by its
swollen club-shaped antennae, the apical much wider than the basal segments.

In South West Africa two genera are most commonly met with, each represented by a single species, *Aspidopleres intercalatus* and *Diphtherogaster
flavus*. These are typical of hot semi-arid regions and are distinguished by
their large size and robust bodies, consisting of well over 100 segments; though
limited to the western half of the continent their distribution extends from
the south-west Cape to Angola.

*The Scolopendromorpha* — A key to the South African genera

1. Small centipedes, without eyes; tarsi of legs with 1 segment; always
   yellow-brown in colour .................. *Cryptops*
— Larger centipedes, with 4 small eyes on each side of the head; tarsi
   of legs always with 2 segments; colour often greenish, occasionally
   blue, yellow, yellow-brown or terra-cotta, sometimes with dark
   cross-bars or stripes ..................................... 2
2. The last pair of legs expanded at their ends into flattened leaf-like
   or racquet-shaped structures, colour orange-red ...*Alipes*
— The last pair of legs like the remaining legs but longer, thicker and
   with more spines; colour olive green, or deep blue-green, or yellow,
   with or without dark green cross-bars ........................ 3
3. Spiracles (in the membranous pleurites at the side of the body)
   oval or round ........................................... 4
— Spiracles longish triangular, parallel to the long axis of the body and
   pointed in front ......................................... 5
4. Size moderate or small, the last pair of legs very slender and almost
   unspined; colour of entire body deep blue-green; the first three
   joints of the antennae smooth, the remaining ones with small fine
   hairs having the appearance of fur ........... *Rhysida*
— Size very large, robust, colour not blue-green, the last pair of legs
   strong in build and well-spined; the first four joints of the antennae
   smooth, the remainder hairy .............. *Ethmostigmus*
5. Antennae very short, not reaching backwards beyond the first segment of the body ....................... *Asanada*
— Antennae longer, reaching backwards well beyond the first segment
   of the body ............................................. 6
6. Second last segment (the first of the two tarsal segments) of all the
   legs, without a spine ..................................... 7
— Second last segment of all the legs, except the last pair, with a small
   spine on its under side ................................... 8
7. The dorsal plate of the first body segment not overlapping the head
   .................................... *Colobopleurus*
— The dorsal plate of the first body segment overlapping the hind
   margin of the head, usually light bronze green in colour ........
   .................................... *Cormocephalus*
8. Head free, meeting the first dorsal plate or overlapped by it ......
   .................................... *Trachycormocephalus*
— Head overlapping the first dorsal plate ...................... 9
9. No spinules at the base of the claws of the last legs; colour dull red
   with a blue or purplish tinge; small to moderate in size; a single
   species, *formosus,* South West Cape .......... *Arthrorhabdus*

– A pair of spinules at the base of the claws of the last legs, colour green or yellow, usually with a darker cross-bar on each body segment or body deep black, the legs bright orange; large and robust centipedes
. . . . . . . . . . . . . . . . . . . . . . . . . . . . . . . . . . . . . . *Scolopendra*

*Cormocephalus, Cryptops* and *Scolopendra* are the commonest genera in this order and are found practically everywhere in South Africa but *Scolopendra* is noticeably absent from the districts of the south-west Cape. *Cryptops* and *Cormocephalus* are indicators of indigenous forest while *Scolopendra* is seldom an inhabitant of forested regions but is usually found in open semi-arid country with a rocky or sandy substratum. *Arthrorhabdus* is common in the south-western districts of the Cape Province, especially in the Peninsula, and in Namaqualand, but does not occur elsewhere.

The giant centipede *Ethmostigmus* is a migrant from tropical Africa and is only found in the sub-tropical regions of Southern Africa such as Zimbabwe, northern Transvaal and Maputo; *Asanada* has much the same distribution as *Ethmostigmus* but has a strong proclivity for sandy semi-desert areas. *Colobopleurus* is not common and is limited to the drier parts of South Africa with a restricted rainfall, such as the Great and Little Karoo. *Trachycormocephalus* is another genus which is more an inhabitant of the tropics than temperate South Africa although it is found in its northern areas and in the border states north and east of the Republic; three species live in the northern half of South West Africa.

The relationships of the South African Scolopendromorph fauna are in general with the rest of Africa. Of the ten resident genera, only four are shared with Madagascar.

Four of the ten South African genera are represented by only a single species, *Scolopendra, Arthrorhabdus, Asanada* and *Ethmostigmus. Cryptops* with 12 and *Cormocephalus* with 23 are genera with a comparatively large number of species.

*The Lithobiomorpha* – A key to the South African genera

1. Genital appendages of the male with 1 or 2 joints . . . . . . . . . . . . . .  2
– Genital appendages of the male with 4 joints . . . . . . . . . . . . . . . . . .  3
2. Antennae with 26 or 28 joints . . . . . . . . . . . . . .  *Walesobius*
– Antennae with 50 joints . . . . . . . . . . . . . . . . . .  *Lithobius*
3. Antennae with 13-17 joints . . . . . . . . . . . . . . .  *Anopsobius*
– Antennae with 19 or more joints . . . . . . . . . . . . . . . . . . . . . . . . . . .  4
4. Basal joint of genital appendages of the female with 5-6 saw-like teeth . . . . . . . . . . . . . . . . . . . . . . . . . . . . . . . . . . . .  *Lamyctopristus*
– Basal joint of genital appendages of the female with 2 or 3 style-shaped spurs . . . . . . . . . . . . . . . . . . . . . . . . . . . . . . . . . . . . . . . . . . .  5

5. Tarsi of legs 1-12 with 2 joints . . . . . . . . . . . . . . *Paralamyctes*
  — Tarsi of legs 1-12 with 1 joint . . . . . . . . . . . . . . *Lamyctes*

The number of species of Lithobiomorpha is small; they are divided among
six genera, and of these *Lithobius, Anopsobius* and *Walesobius* consist of only
a single species, *Lamyctopristus* of two. The single representative of *Lithobius, L.peregrinus,* has been found only at Cape Town; it is probably a native
of Central Europe and like many other small creatures has no doubt been
introduced into our country by means of shipping. The two remaining genera
of stone-centipedes are larger, *Paralamyctes* containing four or five species and
*Lamyctes* seven.

The largest member of the South African Lithobiomorpha is *Paralamyctes
spenceri,* a reddish-brown shiny species; it is the lithobiid that a collector will
be bound to come across, being by far the most common representative of
this group; it is generally found in forest litter or in crevices on the underside
of a damp, decaying log in all the coastal and montane forests from the Cape
Peninsula to Swaziland and probably beyond, with a vertical distribution
from sea-level to 9000 ft (2 750 m).

The species of *Lamyctes* are much smaller and more prevalent in rocky
and semi-arid regions; the only species of lithobiids from Namaqualand and
S.W.Africa belong to this genus. *L.africana* and *L.castanea* have almost certainly been introduced as they occur abundantly in the neighbourhood of sea-
ports and in plantations of exotic trees; to this list of imports the single
species of *Anopsobius* should probably be added. The peculiar genus *Lamyctopristus* with its two species is known only from the south-west Cape.

The Lithobiomorpha are among the few centipedes which have been able
to adapt themselves to life in underground caves. Some specimens of *Lamyctes*
have been discovered in the Wynberg caves at the top of Table Mountain but
these were not true cave dwellers living in permanent darkness with the unmis-
takeable hall-marks of cave-life, loss of all colouring, blindness, and lengthen-
ing of the limbs and antennae, but rather troglophiles, living in the twilight
regions near the entrance to the caves.

The South African Lithobiomorpha have been very sporadically collected
and much of the identified material has been based on specimens collected in
a small area, chiefly the south-west Cape, especially the Cape Peninsula. A.
great deal of intensive systematic collecting still requires to be carried out
before we have anything like a clear picture of the extent of our native fauna.

*The Scutigeromorpha* — A key to the South African genera

There are only two genera in South Africa, each represented by a single species; they can be distinguished as follows:

The joints of the legs, excepting the whip-like many-jointed tarsi, with sharp longitudinal keels along their edges; shields of the back with spines
............................................ *Scutigera*
(one species, *S.coleoptrata natalensis*)
The joints of the legs without sharp longitudinal keels along their edges; shields of the back without true spines but with spine-like bristles . . . . .
............................................ *Scutigerina*
(one species, *S. weberi*)

*Scutigera coleoptrata natalensis* is probably only a local variation of the species *Scutigera coleoptrata* living in many different parts of the world; it is found all over South Africa and is most probably an introduced form. *Scutigerina weberi* is also extremely widespread, especially in the forests of South Africa, but is perhaps an indigenous species.

# VI. THE LITERATURE

It is unfortunate that there are extremely few treatises on centipedes written in simple language that can be understood by the average educated person; on the other hand there is a wealth of interesting information of a detailed nature which has to be gathered from various papers in a large number of scientific journals.

Most of the monographs in which this class of animals is comprehensively treated are either in French or German and there is a great lack of general works in English. One of the older and more helpful accounts is the synoptic article by R.I.Pocock in the 11th edition (1901) of the Encyclopaedia Brittanica, though by most people it would be considered too technical to rate as a popular account.

The German zoologists are the acknolwedged authorities in this field and two of them, C.Attems and K.W.Verhoeff have contributed monumental works on centipedes. These however are too advanced to be of much assistance to any but the specialist and some of the information they contain has now been superseded.

A great deal that is of interest concerning the habits, life histories, development and distribution of centipedes can be gleaned from the volumes of the two above-mentioned authors if the reader is prepared to wrestle with scientific terminology of a rather formidable nature in the German language; nearly all of it is directly applicable to South African centipedes since the four main groups of these animals are found in all parts of the world and differ only in detail from one region to another.

Other general works dealing with Myriapoda as a whole, either in the form of substantial articles, chapters in monographs, or text-books are: R.E.Snodgrass on arthropod anatomy (1952) and the chapter on Myriapoda in A.Kaestners *Invertebrate Zoology* of which there is an English translation by H.W. and L.R.Levi (1968).

A quite recent and very important treatment of the Myriapoda is embodied in *The Arthropoda* (1977) by S.M.Manton; it includes a detailed study of the morphology, function and habits of the group with new research into muscular anatomy and locomotory mechanisms of all four orders of centipedes. The most recent arrival in time (1980) is J.M.Demange's *Les Mille-pattes* written by one of the worlds foremost myriapodologists with many years of field and museum experience. It is however mainly concerned with the European faunas.

Turning to papers and articles which confine themselves to one special aspect of the subject, a quite detailed account of the mating habits and reproductive organs of centipedes was written as long ago as 1855 by Professor L. Fabre (not the celebrated French naturalist and writer, J.H.Fabre). Though some points in his work have been corrected and much detailed information supplemented by later writers, Fabre's main contribution stands the test of time amazingly well. It is lamentable that copies of this work are now extremely difficult to obtain and are not to be found in any of the South African libraries.

Another useful paper in Germany by R.Heymons (1901) gives a good and well-illustrated account of the development of *Scolopendra*, a centipede common in South Africa.

A short paper (1947) on the development from egg to half-grown (adolescent) stage of the Natal forest centipede has been contributed by Lawrence; some of the various stages are illustrated by six good photographs of living specimens.

H.K.Brade-Birks (1920) has given a historical and anatomical account of the light producing organs in the earth-centipedes (Geophilomorpha).

The systematics of the South African Chilopoda have been much neglected and it is only recently that some attention has been paid to them. In 1928 Carl Attems of the Natural History Museum in Vienna, published his monograph on the systematics of the South African fauna. Apart from the monographs of Attems and Verhoeff referred to above what little has been accomplished in the field of taxonomic zoology is represented by the papers of the present writer. In some of these, an attempt has been made to give as complete a list as possible of the faunas of various well-demarcated regions of South Africa, as for example Natal, the Kruger National Park and South West Africa, but only one, (in *South African Animal Life,* 1955), for the South African fauna as a whole.

Of special interest to the South African naturalist must be Barbara Brunhuber's *The mode of reproduction in the centipede,* a degree thesis based on an investigation of the Cape centipede, *Cormocephalus anceps.* It is the only work which deals fully and for the first time with courtship, mating and development in a South African myriopod. It is to some extent a comparative study of reproduction in all the four orders of Chilopoda and sets a high standard of research embodying the results of patient observation over a period of months, if not years. Mating behaviour in any degree of completeness has only been described once before, by Klingel in 1960, for a species of centipede not found in South Africa and Brunhuber's study supplies a number of details on aspects which were previously unknown or not completely understood. Her finding of microscopic pore canals in the envelope of the spermatophore is a new discovery and may go far to explain how the spermatozoa are released in the scolopendromorph centipede.

Four important papers by Klingel of Germany and five by Demange of France dealing with the same subject are extremely valuable and stimulating contributions and as essential reading will be found in the biographical list.

# VII. A SELECTED LIST OF WORKS ON CENTIPEDES IN GENERAL AND THOSE DEALING PARTICULARLY WITH THE SOUTH AFRICAN FAUNA

Attems, C. 1926. Chilopoda, in *Kükenthal's Handbuch der Zoologie*, 4:239-402.

Attems, C. 1928. The Myriapoda of South Africa. *Ann. S.Afr. Mus.*, 26.

Attems, C. 1929. Geophilomorpha, in *Das Tierreich*, 52.

Attems, C. 1930. Scolopendromorpha, in *Das Tierreich*, 54.

Attems, C. 1934. The Myriapoda of Natal. *Ann. Natal Mus.*, 7.

Brade-Birks, H.K. 1920. Luminous Chilopoda. *Ann. Mag. Nat. Hist.*, (9) 5.

Brunhuber, B. 1967. The mode of reproduction in the centipede *Cormocephalus anceps anceps* Porat. MSc degree thesis, University of Cape Town.

Demange, J.M. 1980. Les Mille-pattes. *Boubée, Paris.*

Fabre, L. 1855. Recherches sur l'anatomie des organes reproducteurs et sur le developpement des Myriapodes. *Ann. Sci. nat.* (ser.4), 3.

Heymons, R. 1901. Die Entwicklung der Skolopender. *Zoologica*, 13(3).

Kaestner, A. 1969. Myriapoda, in *Invertebrate Zoology*, 2.

Kitching, J.W. 1980. On some fossil arthropoda from the Limeworks, Makapansgat Potgietersrus. *Palaeont. Afr.* 23:63-68.

Klingel, H. 1956. Indirekter Spermatophoren-Übertragung bei Chilopoden beobachtet bei der 'Spinnenassel' *Scutigera coleoptrata. Naturwissenschaften*, 43.

Klingel, H. 1969. Indirekter Spermatophoren-Übertragung bei Geophiliden (Hundertfüssler, Chilopoden). *Naturwissenschaften*, 22.

Kraepelin, K. 1903. Revision der Scolopendriden. *Mitt. Naturhist. Mus. Hamburg*, 20.

Lawrence, R.F. 1947. The post-natal development of the Natal forest centipede, *Cormocephalus multispinus. Ann. Natal Mus.*, 11(1).

Lawrence, R.F. 1949. The young of Centipedes. *Illustrated London News*, 216 (5795).

Lawrence, R.F. 1953. The biology of the Cryptic fauna of forests. A.A.Balkema, Cape Town.

Lawrence, R.F. 1955. Chilopoda, in *S.Afr. Anim. Life*, 2.

Lawrence, R.F. 1966. The Myriapoda of the Kruger National Park, *Zoologia Africana*, 2(2).

Lawrence, R.F. 1975. The Chilopoda of South West Africa. *Cimbebasia.* Ser. A, 4(2).

Manton, S.M. 1977. The Arthropoda. *Oxford University Press.*
Porat, C.O. 1871. Myriapoda Africae australis. I. Chilopoda. *Öfvers Vet. Akad. Forhandl.,* no.9.28.
Pocock, R.I. 1901. Chilopoda, in *Encyclopaedia Brittanica* (11th Edition).
Scudder, S.H. 1890. On the Myriapoda from the coal measures of Mazon creek. *Mem. Boston Soc. Nat. Hist.,* 4.
Snodgrass, R.E. 1952. A text-book of Arthropod Anatomy. Comstock Publ. Assoc. Ithaca.
Verhoeff, K.W. 1900-28. Chilopoden, in *Bronn's Ord. u. Klas. des Tierreich,* 25.

# VIII. THE MILLIPEDES (DIPLOPODA)
# IN GENERAL

## 1. *An introduction to this class of Myriapoda*

The ordinary and familiar millipede, known to everybody in South Africa as the 'Songololo', is not an exciting creature to look at; in fact it is rather a drab animal, dull brown or blackish, very seldom relieved by rows of spots or banded with stripes of a bright colour; slow-moving and apparently unperturbed as it makes its 'slow but sure' progress to wherever it is going, as if sure that its numerous legs, though individually weak, will bring it safely to journeys end in their own good time. If picked up its reactions are mild and unexciting compared with those of most spiders, scorpions or insects; it merely curls itself in a clock-spring spiral and remains there motionless.

It is as if it knew it belonged to the damp earthy places, the wet soil and sunless forest floors which reek of decaying leaves and shreds of rotting bark, half covered with bright fungus growths. For this is where they are most often found, or, if woodlands are not available, as is so often the case in our drought-stricken land, then a stone, a fallen aloe stem, a deserted termite mound or whatever other debris may be lying about will serve as a shelter against the heat and aridity.

It would however be a mistake to imagine that all millipedes are as unadventurous as they appear; there is always the exception to prove the rule. Taking one example, while conceding that millipedes are colourless vegetarians existing on a flavourless diet of decaying leaves with a certain amount of gritty soil to go with it for 'afters', there is an interesting exception, by name *Callipus*, living in North Italy, which though an honest millipede, is actually a flesh eater. It is a scavenger which has developed carnivorous habits and the shortened front pair of legs are armed with a comb of strong spines for holding the food while running with the other, longer legs. In the process of evolving these carnivorous habits, it has also developed agility and speed, being able to climb rock faces at almost any angle in pursuit of prey and to run almost twice as fast (5,5 cm per second) as the average worm-like (juliform) millipede of comparable size.

Few people have heard of a jumping millipede but there is one, *Diopsiulus*, living in Sierre Leone which can jump forwards to a distance of 2 to 3 cm; when disturbed it performs a series of hops, probably as an escape reaction. Unfortunately we have neither a carnivorous nor a jumping millipede in South Africa but there are others which are almost as interesting. Among them are the brilliantly coloured spirobolid millipedes of medium size which live in the fringe of coastal forest ranging from the Cape Peninsula via Natal to tropical East Africa. Some are excellent climbers of small trees and shrubs while others may live in the leaf mould of the forest floor. The outstanding feature of these millipedes is a brilliant red colouring, their bodies shining like sticks of fresh sealing wax, sometimes ringed by alternating black bands or with an elegant pattern of black spots. They are all aposematic animals displaying a bright red warning colouration and when handled quickly secrete a brown liquid which oozes from minute pores at the side of the body. There are numerous species, some more brightly coloured than others and among them one or two are able to eject very fine jets of the odoriferous defensive fluid from one of the pores situated near the head. The substance has some of the properties of iodine and like it stains the fingers and produces a burning pain if received in a cut finger or in the eyes. A spirobolid, *Rhinocricus,* living in North America, has almost exactly similar habits to those of our South African millipede which has received the name *Chersastus.*

Among our smaller millipedes, the members of the family Odontopygidae have defensive reactions quite different from those of the much larger and less active spirostreptids, i.e. those millipedes which are generally referred to as 'Songololos'. When handled, these little millipedes are extremely lively and react with violent serpentine contortions of the body while trying to escape. The small creatures even try to nibble at the fingers which are holding them.

If released they execute a manoeuvre which seems peculiar to the millipedes of this family; they quickly turn over on their back and with vigorous S-like motions of the body travel quite fast for a distance of about 30 cm, thus not using their small legs at all; having reached a safe distance they turn over again and walk in the normal way, an escape reaction which is quite different from that of any of the other South African millipedes.

Of all the multilegged invertebrates that walk the earth, the millipedes or Diplopoda have by far the largest number of legs although this can vary greatly from 13 pairs in the tiny 'pincushion' millipede *Polyxenus,* to several hundred among certain of the so-called 'sucking' millipedes or Colobognatha. South Africa probably holds a world record in numbers of legs in *Nematozonium longissimum,* a Colobognath living in the forested foot-hills of the Drakensberg mountains in Natal. This elegant but extremely attenuated creature, as is implied by its name, has no less than 355 pairs of legs.

The millipedes have already been compared in a general way with centipedes and the differences which separate these two groups will become more clear in the following account of the millipedes as a whole.


2. *The duplicated body rings*

The skeleton of millipedes is formed chiefly of the hard chalky substance, calcium, and the white bleached remains of dead specimens which we frequently find lying on the open veld show that the armoured covering of these animals is composed chiefly of lime. While these brittle remains break easily in the hand and can be rubbed into a powder as in the case of many sea shells, the body rings of a freshly killed specimen become pliant when treated with a weak solution of hydrochloric acid. Millipedes have an almost rigid and unbendable skeleton and they are therefore unable when walking to make those snake-like wriggling movements which centipedes with their more flexible chitinous bodies are able to perform; on the other hand the millipedes are able to roll themselves into a fairly tight close-fitting spiral which centipedes never do. Each body ring is made up of two pieces, a solid, almost completely circular ring, and on the underside of the body a small plate, the floor or sternum to which the legs are joined and where the spiracles, the slit-like openings of the breathing organs, are found. These two parts are firmly fused together and to all intents and purposes form a single solid ring.

Each body ring in the millipede is really a double ring, the diplosomite, and this term and the name diplopod (double-footed) really refer to the same thing. The millipede body ring consists of two rings which in the remote evolutionary past were separate segments that have become fused; the line of fusion can be seen as a fine suture about half way between the anterior and posterior margins of the segment. The diplopod body segment is thus a duplicated or double one; there is two of everything — two pairs of legs, two of spiracles and internally two pairs of ostia in the heart, and two pairs of ganglia in the nerve cord for each body ring.

Each body segment can be likened to a piece of short piping with the hind end a little widened so as to fit loosely over the front end of the segment behind it, in the same way that large earthenware drainpipes are fitted into each other; each ring is further connected with its neighbour by a tough and slightly elastic membrane which allows a small amount of movement between the two segments (Figure 20a, b). This, however, does not explain how the millipede is able to roll up in a spiral, for obviously a drainpipe could never be made to do this. The millipede can form a spiral because each segment is a little longer along the upper side than along the under side to which the feet are attached. When the animal is disturbed and wishes to enroll it falls quickly on its side and from this position rolls into a spiral; in this attitude the legs and softer under parts of the body come to lie on the inside and are protected by the hard upper side of the body, which then forms the outer boundary of the spiral.

This device of rolling up into a spiral or a ball when danger threatens is practised by all millipedes except the small pin-cushion millipede, *Polyxenus* and is one of the most conspicuous and characteristic postures of the group, distinguishing them at once from centipedes. Many other lower animals practise this defensive attitude in the same way as the millipedes, among them being woodlice (crustacea), some cockroaches, some mites and a rather specialised family of minute scarab beetles found in the rain forests of Natal; the

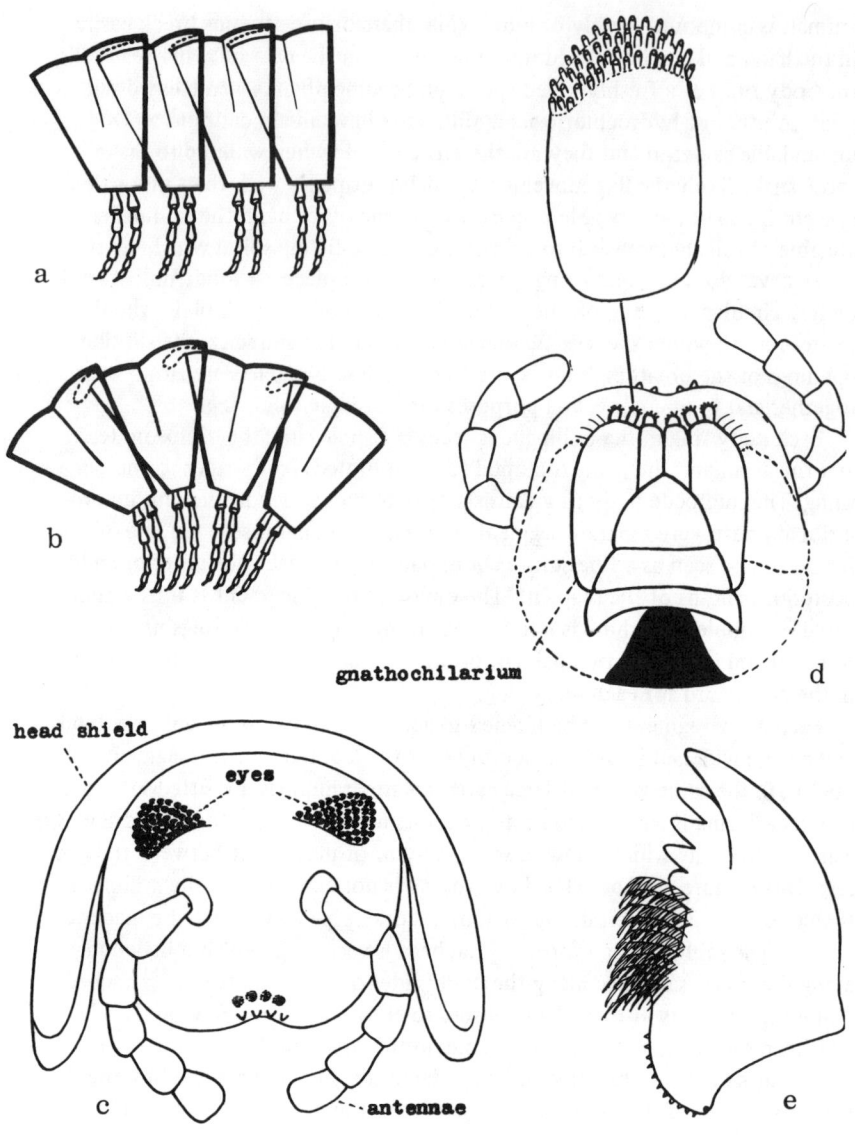

Figure 20. a, four body rings in a juliform millipede in normal position; b, the same when enrolled; c, the head from in front; d, the head from below showing the chin-plate (gnathochilarium), lower part of mandibles (dotted lines), and a taste lobe of the chin-plate enlarged; e, biting portion of the mandible.

68

higher animals also provide good examples of the habit, such spiny or strongly armoured creatures as the hedgehogs and armadillos. Among animals which have long since passed away from the earthly scene, the ancient trilobites, the forerunners of all the animals with jointed bodies and hard outer skeletons, particularly favoured this method of defying their enemies, the more so as they were provided with a strongly armoured body sometimes covered with numerous formidable spines or teeth.

## 3. *The mouthparts*

The mouthparts of millipedes, which to all intents and purposes are all inoffensive vegetarians, are much simpler than those of the centipedes and are well adapted for dealing with the particular kind of food which these creatures favour, such as rotting leaves and wood made soft by prolonged damp and exposure. Less often do they live on the leaves of fresh and living plants and with the exception of one European species they never prey on or eat the flesh of other animals. The mouthparts of millipedes consist of only two elements and one of them, the pair of mandibles, is armed with blunt and rather clumsy teeth (Figures 20d, e); these are to some extent able to break up and grind the other rather coarse food into smaller particles for swallowing; the rest of the mouthparts used for dealing with food are combined in a single large plate which forms the floor of the mouth; this chin-like structure, the gnathochilarium (Figures 20d, 21c), plays little part in chewing but bears on its margin six large bud-like sense organs resembling the organs of taste in the higher animals (Figure 21c). This large plate can be clearly seen on the underside of the head; on each side of it lies one of the mandibles but only the bottom part of this structure can be seen; its apex, which is armed with blunt teeth, setae or spines, lies hidden beneath the chin plate.

## 4. *The senses and sense organs*

What can be said of the senses of the millipedes? The antennae or feelers are very much shorter than those of centipedes and in all the members of the numerous orders of Diplopoda are always composed of eight segments; this number can almost be called a hall-mark of the millipede class. Though only six segments can be counted, the two end segments being extremely small, the seventh is just visible while the eighth is sunk or telescoped into the seventh (Figure 21a). The antennae must be important as organs of exploration, for as the millipede walks along the tips of the antennae are constantly tapping against the ground. It is not surprising therefore that the various segments bear a number of hairs sensitive to touch, but the most important organ found on the antennae is the organ of smell; this nearly always consists of four large cones placed side by side in a group at the extreme tip of the last joint (Figure 21a).

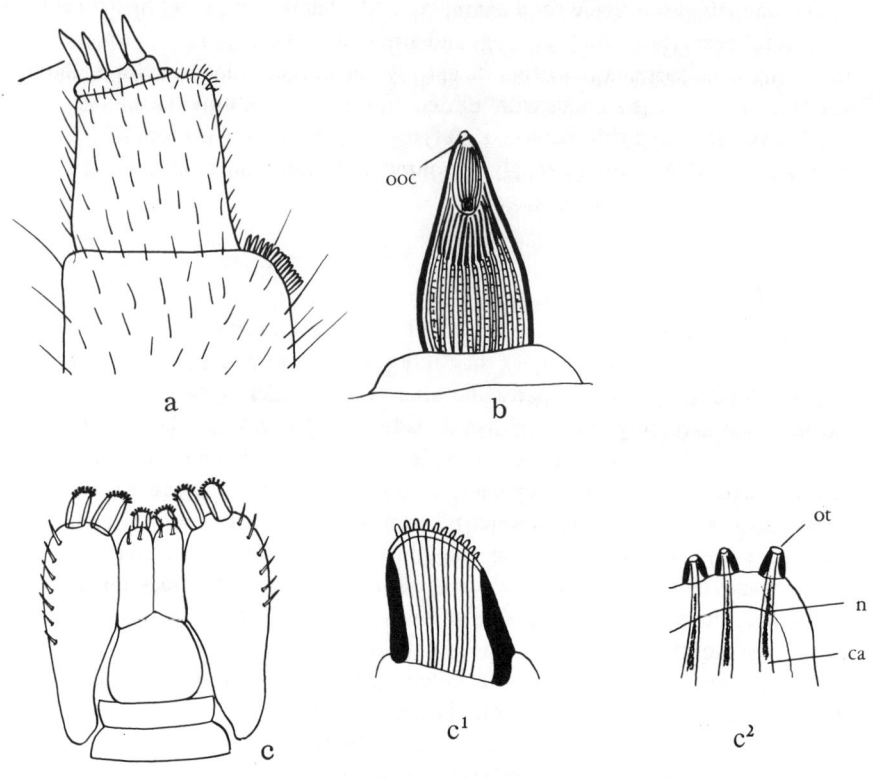

Figure 21. a, the four sense cones at the top of the antenna of a keeled millipede; b, the olfactory sense cone of a pill millipede enlarged, showing the opening at the apex (*ooc*); c, the generalised gnathochilarium of a juliform millipede to show the taste lobes; $c^1$, an outer lobe enlarged, $c^2$, part of the same with three papillae further enlarged.

This sense organ has, as in the case of the centipede, to come into actual contact with the object in question and is perhaps better able to identify liquid or semi-liquid substances than solids (Figure 21b). These four sense cones are also characteristic of the Diplopoda, being found in all the members of this very large class of Myriapoda.

The eyes of millipedes vary a good deal; in the more well-known groups such as the large worm-like millipedes or 'songololas' there is a cluster of numerous close-set ocelli arranged in a triangular group just above the antennae (Figure 20c). These little eyes have very flat lenses and thus do not stand out from the head; neither do they differ much from it in colour, so that they are somewhat difficult to distinguish from the surrounding cuticle. In outward appearance the eyes of millipedes superficially resemble the closely compressed, compound or faceted eyes of insects, with their numerous lenses; though fairly numerous (4-90) these ocelli are by no means as well developed as the eyes of insects and sight in the millipedes must be regarded as poor and not of much importance to the animal in finding its way about or discovering the whereabouts of food. One very large order of millipedes, the keeled millipedes (Polydesmoidea), has no eyes at all and in another there are usually only four small well separated simple eyes or ocelli.

Touch and smell, the most highly developed senses of the millipede, are closely allied; a sense of hearing is not known among them except in the case of one group which will be mentioned later. In addition to these senses the millipede probably has some means of becoming aware of changes in the surrounding temperature of the air and also the amount of moisture which it contains; biologists however have not yet been able to point to any special organ which plays such a part, though it must be one of the utmost importance in the everyday life of the creature.

The organs of taste in millipedes are found along the front edge of the plate which forms the floor of the mouth, the gnathochilarium, and they thus come into direct contact with food which passes into the mouth. Six bud-like structures, arranged in pairs, project forwards beyond the rim of the gnathochilarium and the organs of taste themselves, in the form of numerous little conical papillae, are clustered at their apices, each connected by a strand of nerve fibre to the brain (Figures 21c, c1, c2).

## 5. *The organs of respiration*

The breathing organs of the millipede are situated on the underside of the body near the bases of the legs; there are two breathing openings or spiracles to each segment just as there are two pairs of legs, which underlines the fact that originally each body ring consisted of two separate segments which have become fused into one. These spiracles are very small and slit-like and rather difficult to see; they are situated at the side of the roughly triangular sternal plate and lead into a small chamber which at its end is provided with a bushy tuft of minute thread-like tracheae or breathing tubes which convey the air to the various parts of the segment.

## 6. *Walking and burrowing*

The legs of millipedes are rather weak and are not formed so as to be able to lift the clumsy body far off the ground (Figure 22). Each of these many legs is built on the same plan and they extend sideways when the creature is walking. In order to enable the millipede to walk at even the slow rate it does it is quite obvious that the co-operation of numerous legs is needed for the task and the large number to some extent makes up for the weakness of the indi-

71

vidual members. Millipedes walk in quite a different way to centipedes; the legs have to do all the work of carrying the body and there are none of the supple wriggling movements of the body itself which assist the progress of the centipede; instead the millipede walks straight forward with a peculiar, smooth, deliberate and almost gliding motion.

In millipedes, with two pairs of legs to each body segment instead of one pair as in centipedes, even the smaller kinds have a large number of legs. In the largest forms of juliform or worm-like millipedes that live in the tropics, such as the spirostreptids, there may be as many as 70 pairs or a total of 140 legs, in the smallest about 40 pairs.

After the first summer rains in Natal the spirostreptid *Gymnostreptus pyrocephalus* may appear rather suddenly and in a few days large numbers can be counted. After a few weeks they disappear with equal suddenness. In 1951 such an appearance just outside Pietermaritzburg was welcomed as an opportunity for collecting large numbers of one species to investigate the constancy or otherwise of the leg numbers in a juliform millipede. To begin with 500 were collected at random to establish the sex ratio which turned out to be just one male for every two females; a further 500 were collected selectively to make up equal numbers of the sexes. The variation in the number of leg pairs was 85-99 in the male, 89-101 in the female, thus a variation of 12 or 14 leg pairs with very little difference between the sexes.

If we look at a millipede from the side (Figure 22), or from underneath as it walks over a sheet of glass (Figure 1), we see that the legs do not all move forward together but in groups of 5 or 6 pairs of legs. The movement of the legs thus takes the form of a series of waves like those made by wind blowing over a cornfield, and five or six of them can be seen in action at any one moment of time between the front and hind end of the body; as each leg of a pair acts in exact co-ordination with its partner on the opposite side the waves will be travelling along both sides of the body. The wave motion of the legs of millipedes is very similar to the wave motion of the cilia which can be seen on the bodies of many protozoa and other lowly animals, or on the gills of the common mussel; this beating of a number of cilia in rhythmical sequence is called 'metachronal rhythm'.

In the terrestrial flat-worms (Planaria), and the slugs and snails, which travel along a mucus covered track by means of the 'foot', a narrow muscular strip along the undersurface of the body, an effect rather similar to the wave motion of the legs of millipedes is produced; when these animals walk a number of waves of muscular contraction follow hard upon each other along the sole of the foot which is in contact with the ground, and the animals are propelled smoothly and continuously forward like small caterpillar tractors. In the case of the slugs and flat-worms however the direction of the waves is reversed and instead of travelling from tail to head as in millipedes, they pass from head to tail.

The burrowings of millipedes are of necessity shallow and rudimentary and cannot be compared with the tunnels and runways of such small burrowing animals as mole crickets, trap-door spiders or scorpions which have structures, generally part of an appendage, that can be used effectively for scraping and digging; some species of the Cape scorpion *Opisthophthalmus* for instance, can dig almost vertical burrows in hard clayey soil to a depth of more than a metre. To burrow in this manner would be impossible for even the largest spirostreptid millipede in our country, lacking as it does the necessary digging structures for this purpose. Many of them live on the sun-baked flats of the hinterland where they have to find natural shelters — stones, trunks of fallen trees and sometimes vacated termite mounds where they aestivate in the hot summers, or the unused burrows of other small animals; often in such places there is soft friable soil and they can burrow a little deeper. Some of the larger juliform millipedes can to a limited extent burrow in the leafy humus of our rain forests or in friable soil at somewhat greater depth, bulldozing their way by shoving with their large rounded heads bent downwards.

Millipedes are able to exert considerable pushing motive force derived from the many moving legs and which is transferred to the head end; the weight of a large millipede's body could also be used to hollow out and smooth off a more or less rounded shelter to which it could retire at need, as for instance at the onset of the moulting period.

Some millipedes, especially those with narrow heads like the keeled Polydesmoidea can use them as a wedge to gain entry to a crevice and then by repeated pushing, widen it. Those Polydesmoidea that have more or less flat backs due to the large projecting keels at the side of the body (Figure 28a), which also cover and protect the delicate legs, are good at widening fissures which tend to split along one plane such as occur under bark or compressed heaps of decaying leaves.

Dr S.Manton in her laboratory at London University devised a most ingenious series of experiments for measuring the pushing and pulling power of millipedes by harnessing them to small sledges or pans carrying weights. She found that most millipedes refused to push with their heads but when harnessed by a small strap to sledges of suitable size, made the most willing carthorses.

It was found quite easy to measure the force which could be exerted by a fair sized millipede; this was arrived at by adding weights to the pan as it was pulled over a 'course' composed of rough damp paper; the actual force was calculated by the pan, loaded up to its maximum, being connected by a fine thread over a pulley to another pan hanging freely to which gram weights were then added just sufficient to shift the 'cart'. If a mouse with a body weight equal to one of the large juliform millipedes was harnessed to a 'cart' it was found to exert only a quarter of the pulling power of a millipede. It would seem that there are some fringe benefits for animals having many legs.

## 7. *How to distinguish the sexes*

As has been mentioned in a previous chapter, the millipedes are peculiar and differ from many other invertebrates such as centipedes and insects in having the openings of the reproductive organs placed far forwards, close behind the head, instead of at the extreme hind end of the body. For this reason they have been assigned by zoologists to a group called the Progoneata as opposed to the Opisthogoneata (reproductive-organs-at-the-rear) such as the centipedes. They open on the third trunk segment in both sexes although the external male sex-organs, the gonopods (meaning 'sex-legs'), are found on the seventh segment of the body (Figure 23). These are really ordinary walking legs which have undergone a drastic change of structure and function just as the first legs in the centipede have become modified into offensive weapons — the poison-jaws. Their purpose is to take up the sperms from the reproductive organs on the third segment and transfer them to the female during the mating act. In the spiders the pedipalps, used by them as antennae, have become modified for the same intermediary function. In the millipedes however the gonopods would hardly be recognised as legs as they have become very much changed in the course of ages for fulfilling their highly specialised mission. In many of the larger millipedes they are unbelievably intricate with a complicated pattern of membranes, spines and teeth. The whole apparatus serves as a useful means for distinguishing one species from another, for most millipedes of approximately the same size are outwardly very much alike, especially the females of the various kinds. Most of this complex structure is withdrawn into the body and is hidden from view except during the pairing act but a small dark coloured part of it can be seen protruding from the under surface of the body on the seventh segment and this is one way of telling the males from the females. An easier way of recognising the males however is by their feet, most of which have on the underside two small white fleshy pads which look rather like snow-shoes; the females never have these (Figure 23). In the nuptial embrace these pads enable the male to hold the rather smooth body of his partner more firmly and they bring to mind the rough horny growths on the chest and arms of the male frog which appear during the breeding season and aid him in clasping the slippery female. Other means of distinguishing the males from the females are few, and the general appearance of the two sexes is very much alike. As one becomes accustomed to handling large numbers of the creatures however it soon becomes evident that the males are rather more slender than the females; their 'coats' are more glossy and their walking movements are more rapid. If one looks closely there is a slight hump on the back just behind the head where the 'neck' would be expected to be. One is more likely to come across the active and more enterprising males than the females and it is most probable that the majority of millipedes found walking in the open will be males. After the brief afternoon

74

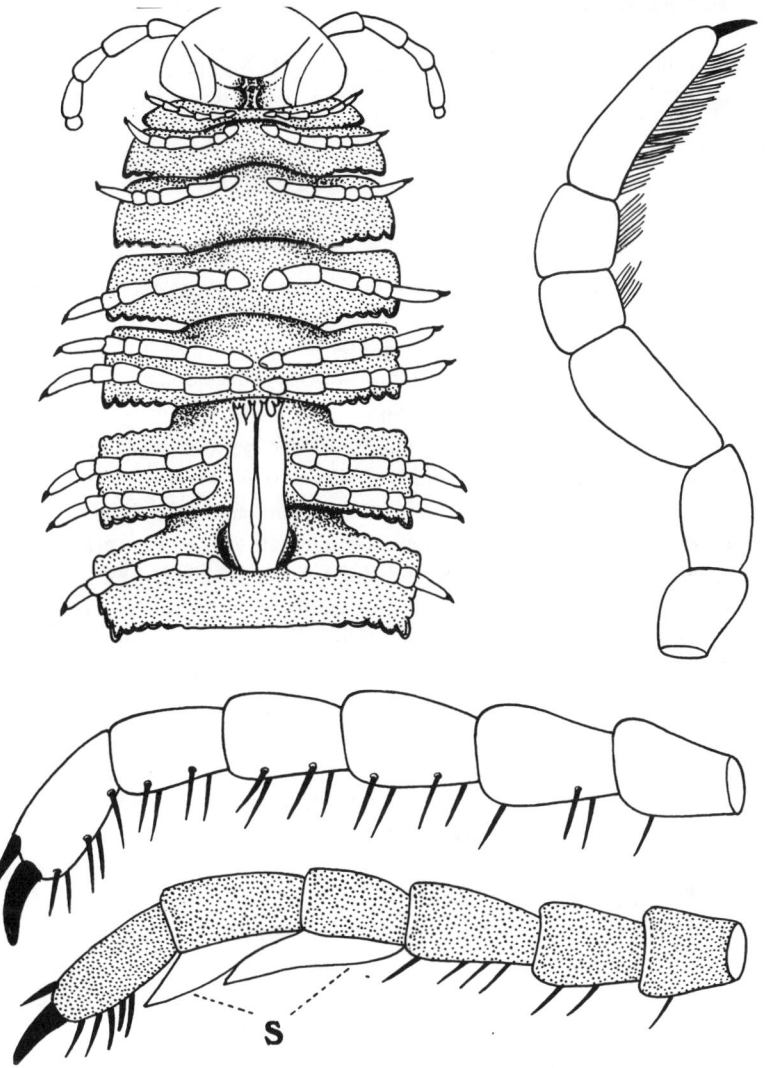

Figure 23. Above, left, underside of the head and first seven segments of a male keeled millipede (*Platytarrus*) to show the position of a male sex organ; above, right, the brushes on the second leg of Platytarrus; below, legs of a female and a male (stippled) of the large spirostreptid millipede, *Doratogonus*.

thunder showers of an up-country summer a familiar sight on the highveld is the enormous spirostreptid millipede, magnificent in his armour of shining rings, moving majestically in the open without fear or concealment. It is the nuptial walk, the *Liebesspaziergang*.

75

In South Africa the spring and early summer months of October to December constitute the mating season but this may be prolonged into March. When the sexes come together the male shows an ardour in his embraces which is surprising and unexpected in creatures whose movements are otherwise so leisurely and unhurried.

## 8. *Mating and transfer of the sperm*

In December 1935 at Pietermaritzburg a number of both sexes of the large black spirostreptid, *Doratogonus setosus* were kept in a glass covered box for observation. It was found easy to induce copulation in these millipedes by merely placing a male near to or in contact with a female; within half an hour all the males had secured females, showing considerable eagerness and avidity and even attempting every now and again to copulate, rather clumsily, with another male; no form of preliminary courtship was ever observed. The male would hold the female firmly by quickly throwing one or two turns of his body around her behind her head (Figure 24), at the same time the protruded gonopods would be applied to the vulva of the female as the two were pressed together, venter to venter, the insertion being accomplished very quickly; almost immediately the whole membranous sac surrounding the base of the gonopods became tumescent, bulging out between the segmental rings. During copula the partners remain very quiet appearing oblivious of other stimuli, the antennae of both trembling a little, while a slight stroking movement by some of the legs is kept up. In this condition they could be lifted up repeatedly by their bodies, placed in sunlight and photographed. When eventually the pair were forcibly separated the male genitalia were seen to be extended to their uttermost but were then partly withdrawn, the tumescence subsiding very quickly. The sex organs of both male and female were seen to be covered with considerable but varying amounts of the milky white sperm fluid. It was now easy to remove some of it with a pipette so as to demonstrate that the large, roughly triangular spermatozoa are not motile. Sometimes if the two partners were allowed to renew contact another copulation ensued; at other times the two would go their different ways.

The sexes having separated, the male almost immediately began to clean his gonopods, taking up a position with the head strongly arched downwards and backwards towards the ventral surface allowing the mouth-parts to come in contact with the sexual areas. The motions of the mouth-parts in this manoeuvre were difficult to distinguish but could best be described as a licking movement, much as a cat licks the fur of its chest but more slowly. In a short time all traces of sperm will have disappeared from the gonopods which were then completely withdrawn. Sometimes the cleaning process was repeated; the female on the other hand showed no inclination to perform a sexual toilet similar to that of the male.

76

## 9. *Egg laying and early development*

Like the centipedes, all millipedes lay eggs, which are small round objects about as large as a pin's head; they are much smaller but far more numerous than in the case of the centipede and several hundreds are usually laid at a time. Many females build a quite elaborate nest in which the eggs are deposited; the nest is made of moist earth voided by the female and worked into shape with the last pair of legs. The eggs are generally laid two at a time into the little cup-shaped nest and the female pauses from time to time to build up the walls a little higher (Figure 25a). The nest eventually assumes a dome shaped appearance (Figures 25b, c, d) with an opening at the top leading into a central chimney for supplying ventilation (Figure 25e). Some species of millipedes even cover the outside of the nest with a rough camouflage consisting of fragments of earth and dirt which renders it practically invisible when resting on the ground; others coil round the nest and guard the eggs with their bodies. Our larger spirostreptid millipedes make rough chambers in rotting wood or damp soil by the movements of their bodies and in these the eggs are deposited. Only one egg is laid at a time and each is protected by a capsule made of earth mould. The procedure of making the capsule is quite an elaborate one and is carried out in the following way. The female lies on her back or side, and taking a small quantity of earth mould, dampens it with saliva from her mouth, kneading it with her jaws, and places it in a flat layer on the under surface of the body between the legs; the egg is then laid upon the flat layer of mould after which the millipede bends the sides of the layer round the egg with her legs and head so as to encircle it completely, forming an enclosed capsule. Finally, she moulds the capsule into a rounded form and finishes it off by rolling it backwards and forwards between her legs as far as the head. At this point she brings the hind extremity of the body into contact with the capsule, grasping it with the anal valves which are stretched widely apart, and in this way covering its whole surface with a special secretion from glands in the intestine.

As many as 300 capsules may be made, each containing an egg; in the largest millipedes they may measure as much as 12 mm long and 8 mm wide but more usually are about 4,5 mm long and 4 mm wide. The wall of the capsule is about a millimetre thick and inside it there is a cavity in which the egg lies free allowing just sufficient room for the newly hatched embryo to move about and feed on the walls of its prison; these serve as its first nourishment during the early stages of development.

The capsules are oval in shape, sometimes a little flattened above and below, and in their general size, colour and external pattern strongly resemble the small pellets of excrement, little heaps of which are often found in the neighbourhood of millipedes. They can be distinguished from them however by the fact that the pellets of excrement are more cylindrical, smaller and slightly flattened at one end.

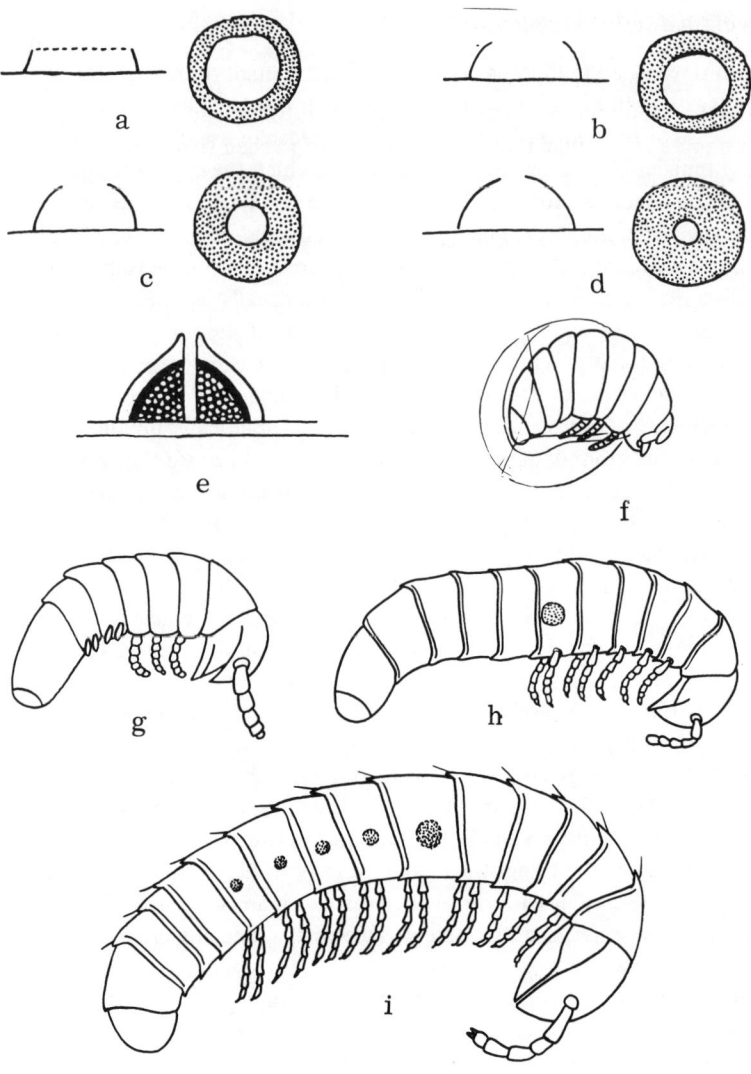

Figure 25. a-e, the building of the egg cacoon of a juliform millipede; f, the six-legged larva emerging from the egg; g-i, further growth stages of the developing millipede.

The egg capsule therefore serves both as food and protection for the young soft-bodied larva during the early and critical stages of its growth but apart from this there is in the great majority of millipedes little sign of that single-minded devotion of the mother for her young which is so striking in the case of the larger centipedes.

The way in which the young develop after they leave the egg is also quite different from what has been seen in the centipede. Instead of hatching from the egg with a large number of crude and stumpy legs, the same number as in the adult, the larval millipede has only three pairs and in this respect it stands on a level with the full-grown insect (Figure 25f, g). In the case of most millipedes the minute larvae take no food at first and soon shed their skins, acquiring more legs and more body rings after each moult (Figure 25h, i). The new segments and legs as they appear are always added at a special point, the growing point, near the hind end of the body. After a while the larvae become more active and begin to feed but they take more than a year to reach the full size of the grown millipede. The young ones are much lighter in colour than their elders, becoming steadily darker as they grow older.

## 10. *The shedding of the skin*

The shedding of the skin goes on not only at regular intervals during the growing stages of the young but also from time to time during the life of the adult millipede. In the full grown animal this is a critical and difficult period since both before and after the moulting of the skin the body becomes soft and movement and feeding are impossible.

Before the creature becomes quite helpless it seeks a convenient shelter or it may even construct an elaborate mud nest similar to the one in which the eggs are laid; hidden in this retreat it passes through the crisis. During the actual moulting or ecdysis even the legs can be of no assistance and the animal lies on its side; the old skin splits cross-ways just behind the head and rhythmic contractions passing from back to front gradually force the long body through this small opening; the old cast skin is otherwise whole and undamaged, though very thin and semitransparent with the glimmering whiteness of a shroud (Figure 24). The process of throwing off the old skin is an exhausting one, making large demands on the creature's resources. It now requires a little time in which to recover itself and for a whole day lies immobile and to all appearances dead; the body which on its emergence from the old cuticle is soft and yielding with the consistency of rubber, slowly hardens once again. In most cases the first act of convalescence is to eat the moulted skin; in this way the valuable supply of calcium stored up in the old cuticle can be used for hardening the new one.

The complicated and difficult process of moulting constitutes a dangerous period in the life of the millipede, the more so since it lasts for so long, about three weeks in all, during one week of which it is helpless and immobile. This is far longer than the period required for moulting in other members of the cryptic forest fauna of comparable size, such as centipedes, insects, woodlice and Peripatus.

## 11. *Defence strategies and the defensive glands*

The natural steps which a millipede takes to protect itself when attacked is
to roll itself into a spiral like a clock spring, with the head, legs and vulnerable
under parts closely pressed together on the inside (Figure 22). Frequently at
the same time it will expel liquid faeces. Certain of the smaller kinds however
disdain this passive attitude and when they are held with the fingers make
vigorous squirming attempts to get away, throwing themselves about with
violent snake-like twisting movements. This behaviour is typical of the small
odontopygid millipedes. It is indeed very difficult to hold their smooth
bodies and they often succeed in escaping, even when grasped with a forceps.
Sometimes the bolder ones even try to bluff that they are more dangerous
than they really are by nibbling at the fingers with their tiny crude jaws, an
astonishing exhibition in such a defenceless animal seeing that the mouth-
parts are really quite unable to do any damage. Others of the spirostreptid
family Harpagophoridae are provided with a sharp tooth-like spike at the tips
of their tails with which they try to probe the fingers, a habit which they have
in common with the little blind and quite defenceless earth-snake *Typhlops*.

Altogether the best defence the millipedes possess is their habit of living
unnoticed in obscure places where they are safe from the attentions of other
carnivorous animals. But they have another weapon to fall back on which
must make them a very unpleasant morsel for any animal which would be so
indiscreet as to take them for a mouthful of food. On nearly all the segments
of the body there is a minute opening or pore about half way up and near the
hind edge of each body ring. These are the openings of the defensive glands,
and in the case of the worm-like millipedes a pair of them is found on each
segment of the body (Figure 29), except the first four or five and the last
two. The glands have been given a number of names such as odoriferous,
repugnatorial, repellant, and less elegantly, stink glands. While the first two
names are long and rather cumbrous the last one is not an accurate description
as the secretions of the glands are pungent rather than offensive or malodorous.
It is perhaps better to call them defensive or repellant as these are the uses to
which they are put. When a millipede is roughly handled a strong smelling
liquid is forced through these tiny apertures in the form of a little drop; it has
an odour recalling both that of iodine and the disinfectant chloride of lime
but somewhat stronger; it has also been described as resembling that of bitter
almonds; actually it contains iodine, hydrocyanic (prussic) acid, and small
amounts of quinine and chlorine, the last of which gives it a characteristic
odour; like iodine it stains the fingers a yellow brown which afterwards turns
to a dull purple colour. If the liquid enters a small cut in the skin it causes a
most painful burning sensation. In some of the tropical millipedes, the liquid
can be squirted some distance from the creature in the form of a fine jet. An
American naturalist, Dr Loomis, tells how in carelessly handling one such

species, *Rhinocricus lethifer,* from the West Indies, he received some of the poison, which was ejected to a distance of about 18 inches, in the eyes; the pain was intense and instantaneous, one eye being closed for several hours; and though bathing with cold water brought relief, blisters on the more sensitive parts of the skin persisted for nearly a week. Mr E.Burtt in East Africa placed a large specimen of *Spirobolus* in his hip pocket and was surprised to find that this caused inflammation of the skin of the hip, the blackened skin of the affected part eventually peeling off and leaving a raw wound. The properties of these secretions of millipedes are evidently familiar to native peoples as they are used by the inhabitants of the Malay Peninsula as an ingredient of the poison with which they coat their arrow heads.

In South Africa the habit of excreting a poisonous or irritating fluid is most highly developed in species belonging to worm-like (julomorph) millipedes of the forest-living genus *Chersastus.* These millipedes nearly all have a highly characteristic vivid red colouring which may be variegated with rows of black or yellow spots. In one species at least, *Chersastus annulatus,* the substance is emitted as a fine jet and when placed on the tongue gives rise to a burning or smarting sensation. This defensive weapon seems to be much more characteristic of the family of smaller juliform millipedes, the Spirobolidae, to which *Chersastus* belongs than of the large black, blackish-brown or reddish-brown millipedes belonging to the family Spirostreptidae; *Doratogonus* for example, which is a member of this family, has much feebler powers of secretion.

## 12. *Some enemies of millipedes*

It is worthwhile noticing that the attacks of assassin-bugs on millipedes, described in the next section of this account, are confined to these larger forms of Spirostreptidae, while *Chersastus* is strictly avoided. The parasitic mites also, which are found in large numbers on *Doratogonus,* are hardly ever known to occur on *Chersastus.* Many observers think that its powerful defensive secretion acts as an antiseptic and bactericide, the chlorine giving it its disinfectant properties; such a function would no doubt be extremely useful to creatures so largely soil dwelling as millipedes and which are always liable to be attacked by fungi and bacteria. In spite of the discouraging effects of the secretion, which in at least some cases must serve as a deterrent to would-be predators, it is surprising how many birds and mammals find it possible to overcome its unpleasant effects and to include millipedes in their bill of fare. The list of birds known to consume them or which have been proven predators after finding them in stomach food counts, is formidable. Nearly all the species known to be consumed by predators, which are almost all larger vertebrate animals, belong to the juliform millipedes and are of middling to large size; practically no observations are on record respecting the other four orders, nearly all small-sized millipedes.

At the present time however any list of the animals known to use millipedes as food would be quite incomplete and would contain only a fraction of the actual number. Most probably many Muridae, mole-rats, and other small burrowing animals feed on the millipedes they encounter in their underground habitat but up to the present no complete census has yet been made of bird, mammal and reptile predators in South Africa. The following list gives but little idea of the actual number.

*Mammals.* In the Transvaal the Cape doormouse, *Graphiurus ocularis,* makes large midden heaps of the disarticulated body rings of the large, widespread spirostreptid *Doratogonus flavifilis* outside its retreat. In the other provinces it feeds on the spirobolid *Chersastus* and the larger *Doratogonus annulipes.*

The mongoose *Suricata* and various genet cats (*Genetta*) together with various other members of the family Viverridae were recorded by Dr Reay Smithers in his book *The Mammals of Botswana* as well as various jackals (Canidae) and two species of pole-cat (Mustelidae).

*Birds.* Francolins, guinea fowl, the hadedah and probably most of the other species of Ibis. Dr J.M.Winterbottom gives the names of the following bird predators: Cape robin, purple heron, little sparrowhawk, common sandpiper, chorister robin and Natal thrush; African robins of the genera *Erythropygia, Cossypha, Bessonornis, Sheppardia, Pogonocichla, Swynnertonia,* the crowned horn-bill *Lophoceros alboterminatus.*

*Amphibia.* Most of the larger frogs and especially toad species of the genus *Bufo,* consume millipedes as well as occasional centipedes when the opportunity offers though none live exclusively on such a diet, as does the small centipede eating snake *Aparallactus* (see p. 19). Inger & Marx (1961) give many examples in the food tables accompanying their paper 'The food of amphibians'.

*Reptiles.* The terrestrial hinged tortoise *Kinixys belliana,* the rock lizard *Cordylus vittiger* and the leguaan *Varanus albigularis.*

*Invertebrate predators.* Apart from the reduviid (assassin) bugs described hereafter in Chapter IX the only known predator is the fairly large rock scorpion, *Opisthacanthus laevipes.* In the Kruger National Park large midden heaps of the bleached skeleton rings of the common Transvaal spirostreptid *Doratogonus flavifilis* (Figure 26), have been found at the mouth of the burrow of this scorpion; it was on one occasion seen holding a millipede in its claws injecting a dose of poison with its sting, quite undeterred by the copious secretions from the defensive glands of the prey.

## 13. *Toilet and grooming habits*

Millipedes both look and actually are very clean creatures; as in the case of the centipedes they spend a good deal of time brushing and polishing themselves with great care and special attention is paid to the gonopods immediately after copulation as has been described on a previous page. At other times the antennae and legs are carefully cleaned by passing them through the mouth where they become covered with moisture provided by the salivary glands. Such secretions probably act as a germicide and disinfectant destroying harmful bacteria from the soil which may cling to the feet of the millipede during its wanderings. The first three legs of the male often carry a special brush or comb of stiff hairs and this useful contrivance is well suited for its purpose of cleaning the antennae.

The thick bush of the Bluff at Durban used to be a paradise for many kinds of millipede and with the aid of binoculars it was fairly easy, even in such a dim twilight, to follow their movements at close hand while occupied with their grooming activities. The species observed was the large black spirostreptid *Doratogonus setosus* which sometimes seemed to use up most of the morning over its elaborate toilet conducted in a leisurely and yet purposeful manner.

All the legs are systematically cleaned and then whatever parts of the body that can be reached with the mouth in much the same way as a cat sitting in the sun licks itself carefully all over. The whole leg from base to apex is cleaned with care, even between the coxae at the base of the limb, the leg being moved first forward and then backward to get at the difficult in-between places; the end segments of the legs are passed through the mouth rather as a Mantis cleans its antennae, each leg being rapidly treated in turn. The genitalia are extruded in a fraction of a second and with complete facility and are then treated in the same manner as the legs, the mouth being bent right down and a little forwards.

It is not surprising that few parasites are found on millipedes although most insects and many spiders and other arachnids of smaller size have mites of various kinds and at different stages attached to their bodies.

## 14. *The absence of social life: Do millipedes migrate?*

Although millipedes live inoffensive lives hidden for most of the time under logs and stones, or in the damp mould which forms a carpet on the floor of our indigenous forests, some of them, especially the larger kinds, can often be seen walking in the open after a summer shower when the air is still laden with damp and the earth is pleasantly moist and cool. They are essentially lovers of moisture and cannot endure dry heat and the unbroken rays of the sun.

As a rule they live unsociable lives each seeking his own sequestered dwelling place; rarely does one come across a pair together under a log or stone. In Europe and the United States however a few species are gregarious and at certain times of the year mass migrations of millipedes have been known to take place; very large numbers swarm like trek locusts, passing across whole districts in spite of the most severe obstacles in their path. In France they have been known to stop a train, the crushed bodies of the victims having made the rails too slippery for the wheels to revolve. Such swarming has never been known to occur in South Africa where the heat of the sun and the general aridity of the air are probably too extreme to make such long migratory journeys possible.

Often after heavy summer ' rainfalls' in the platteland the traveller by car will notice scores of millipedes walking across the road so that for several miles he will not be able to avoid them on the hard macadam and his progress will be littered with crushed corpses. These however do not constitute a migration, a stream progressing steadily in the same direction like the voetgangers and trek locusts, or the millipede swarms of northern countries. The millipedes are taking advantage of favourable conditions to spread at random in all directions from one or more breeding centres; there are such large numbers over an extensive area that while many individuals will happen to cross the road in either direction, there will be just as many walking about in the veld on both sides of the road. A specialist collector of millipedes from the nearest natural history museum will thus be provided with a golden opportunity of obtaining with the minimum of hard work, a fine representative collection for study purposes – if he has the requisite jars and containers handy. In only one species of millipede has a certain degree of gregariousness been noticed in our country. The common and widespread spirostreptid *Gymnostreptus pyrrocephalus* is quite exceptional among South African millipedes for its habit of laying eggs in cattle droppings, where the larvae hatch out; this may account for the unusually wide distribution of the species. When the young have reached a certain stage of their growth and are about a third of the size of the full grown animal, they swarm in large numbers, often lying out in the open sunlight and almost touching each other; for most of the time they lie inert but occasionally one or other among the mass moves a little. The full grown specimens on the other hand are just as solitary and unsociable in their habits as are all the other species which inhabit our country.

## 15. *The distribution of millipedes*

Millipedes in general tend to have a rather localised distribution; individual species seldom occupy a very large lebensraum. Those that are found in many parts of the world all owe their wide distribution to the commercial activities of man, having been casually imported into various countries by shipping together with plants, fruit and often some of the soil accompanying them. Millipedes by their own natural activities can cross water much less easily than spiders or centipedes for instance; they are fairly heavy and sink in most liquids so that rivers constitute almost insuperable barriers for them. As a group they are much more sensitive to their environment than are centipedes, requiring rather special conditions of moisture and a soil which is not too dry and impermeable for burrowing; limitations are also imposed on their geographical dispersal by the long periods required for moulting, during which they are almost completely helpless and at the mercy of their environment.

The millipedes of the cold northern countries are nearly all of insignificant size and it is only in the tropics that the really large specimens of the race flourish. Even in the temperate climate of the Cape Peninsula the largest millipedes are pygmies in size compared with the giants which are to be found in the northern subtropical regions of South Africa, where they may be nearly a foot in length.

Speaking in general terms millipedes will flourish best in regions such as the Amazon and Congo which are largely covered with shady and moist forests. On the flat sun-baked prairies and open steppes, and in arid sandy deserts they will be found, but in smaller numbers. Thus the immense rain forests of the equatorial tropics, with their moist steamy heat and perpetual shade, are a paradise for millipedes, and it is here that they occur in largest numbers, reach their greatest size, and put on their gayest attire.

16. *The part they play in soil formation*

Millipedes, especially the smaller kinds, play an important part in the formation of the rich humus that carpets the floors of our indigenous forests. Together with myriads of other small forest-dwelling animals, such as mites, insects and worms, and such lowly plants as bacteria and fungi, they feed upon and break down the woody debris of the forest. This untidy litter consists for the most part of leaves, seeds, twigs and branches that have fallen to the ground, all in a damp and mouldering condition, and as these fragments gradually break up they become thoroughly mixed with the forest soil. This soil or humus after a time becomes dark and rich in small particles of decaying organic matter, a fertile nursery for the young seedlings that are constantly springing up in the forest to replace the older trees as they die off. Not only do the members of the forest microfauna feed upon this litter but they also continually burrow through it in their comings and goings, turning it over and allowing air and water to circulate within it.

An American biologist, F.H.Colville, has told us that certain forests near Washington in the United States are inhabited by large numbers of a small spirobolid millipede related to our own forest-living *Chersastus;* he calculated that these millipedes, by means of their excrement alone, contribute more than two tons of rich manure in each year for every acre of forest land.

It can thus be appreciated that the work of even such small creatures as millipedes and their forest-living allies is of much importance in the formation of soil and therefore one of direct interest to foresters and farmers.

## 17. *Millipedes as pests*

Millipedes should not be regarded as pests of primary importance and their destructive activities cannot be compared with those of insects, nematode worms, or mites in the realm of agriculture. No parasitic forms are found among them and what harm they do is related to the world of plant and not animal life. In general they prefer already damaged and decaying plant tissues as food and when millipedes are found attacking vegetation this can often be construed as a sympton rather than a cause of damage previously effected by accident or by some more serious pest.

Nevertheless a number of complaints of damage to crops have been filed against various millipedes belonging to two groups, the worm-like Juliformia and the keeled millipedes or *Polydesmoidea,* though it is far more often the *Polydesmoidea* that are the culprits. Such cases are more frequent in well populated countries where market gardening and the intensive cultivation of crops in small holdings is widely practised. Vegetables and flowers raised under roofs and in greenhouses also offer grand opportunities for the onslaught of numerous pests, including millipedes; in gardens they attack a number of plants, usually by destroying the roots or those parts of the plant situated below ground level.

In Europe small millipedes mostly belonging to two genera, *Blaniulus* and *Cylindroiulus* have been known to attack the following plants: potatoes, Brassica, peas, beans, cabbage, sugar-beet, cucumber, swedes, strawberries and a large number of garden and greenhouse flowers. In South America some species of the large spirostreptid millipedes have been accused of damaging potatoes.

In the group of keeled millipedes (Polydesmoidea) some species of *Polydesmus* and *Brachydesmus* are suspected of causing damage to various plants but this reputation does not appear to be founded on very definite evidence. The hothouse millipede, *Orthomorpha gracilis,* which has spread to many tropical and temperate countries of the world, including South Africa, usually or very often occurs in greenhouses; when a foreign imported animal is found in large numbers, as it almost invariably is, it is liable to be labelled as a pest; this is the case with *Orthomorpha* but there is little evidence of damage done since neither the young stages nor the adult feed on green living plants. It has however become a minor pest in the Johannesburg gold mines where it has been found to cause damage to wooden pit props thousands of feet below ground level.

86

In South Africa large keeled millipedes of the genus *Ulodesmus* are often found in gardens and may be suspected of doing a certain amount of damage to crops.

With regard to the other millipede groups, only two come within the orbit of man's activities. The minute pin-cushion millipede *Polyxenus* has been credited with the ability to convey spores of the potato disease by an agricultural authority in Germany. The second case is far more serious and is well authenticated; it concerns *Scutigerella immaculata,* a small myriapod which strictly speaking is not a millipede at all, as it is entitled to a separate class in its own right related to both millipedes and centipedes, the Symphyla. The serious depredations of vast numbers of this small myriapod have become a menace to market gardeners in the United States and parts of Europe such as Portugal. It is difficult to destroy and feeds upon a wide range of edible crops including lettuce, cucumber, radishes, parsley, spinach, celery, carrots, egg plant, beet and asparagus, as well as a number of ornamental plants and flowers.

18. *Numbers and size*

The species of millipedes outnumber those of the centipedes by 4 to 1, and of the approximately 8 000 kinds found in various parts of the world, about 350 have their home in South Africa. They are also a much more complex group than the centipedes and differ far more among themselves in shape, size and general appearance; while it should be easy for anyone to distinguish any of the four groups of centipedes, some millipedes can with difficulty be recognised as belonging to the millipede class at all. With regard to size for instance, the largest species, *Graphidostreptus gigas,* is a little short of 11 inches, the smallest only a fifth of an inch in length; some are long and worm-like, others short and almost round; some have keels or are curiously ornamented with tubercles, spines and hairs, though on the other hand even more have perfectly smooth bodies. With the aid of illustrations however, it should be possible for the enthusiastic student of animal life to distinguish the main groups into which the millipedes are divided, and to learn something of the peculiarities of habits and structure in each.

# IX. THE FIVE SOUTH AFRICAN ORDERS OF MILLIPEDES*

## A. THE WORM-LIKE MILLIPEDES (JULIFORMIA OR HELMINTHOMORPHA)

### 1. *General appearance*

These are just the thousand-footers or 'songololos' with which everyone is familiar. They are elongate worm-like millipedes with a large number of segments — sometimes as many as 70, though the number usually falls between 40 and 60, and almost twice that number of legs. Their bodies as a rule are quite smooth and shiny, with no hairs, spines, or little knob-like outgrowths such as decorate the bodies of the keeled millipedes which will be met with in the next section. In one of the spirostreptid families, the Harpagophoridae, the tail is armed with a strong turned-up spike but in all others this is completely absent. There is always a compact group of numerous little eyes arranged in the form of a triangle at the base of the antennae (Figure 20c). The Juliformia are the commonest of all the millipedes and are what everyone understands by the term; they are the kind most frequently seen in gardens or on the open veld and they also include the largest and most conspicuous members of the class as a whole.

### 2. *Colouring*

The colouring of most of these millipedes is rather dull and they favour a uniform colour which is usually brown or blackish-brown, without decorations in the form of spots or stripes. One or two exceptions however stand out by reason of their unusual and pretty colouring; one of these is the forest-living *Chersastus* which is a very handsome and brilliant red that shines in the sun like fresh sealing-wax; the various species may be ornamented with rows of black or yellow spots which make a very attractive pattern ; another species, found only in Namaqualand, probably a species of *Harpagophora,* is undoubtedly South Africa's most beautifully coloured millipede, for the body is a deep indigo blue, blending into green and yellow just as the colours of the spectrum merge into each other; down the middle of the back there is an amber stripe while the legs and antennae are deep black; altogether it is most unusual for a millipede to be clothed in a coat of many colours like this and the great majority of our species are puritans wearing only a single colour and that an unattractive buff, brown or black. Another exception is the large robust spirostreptid *Doratogonus flavifilis* which is very common in the Kruger

---

* A sixth order, the *Nematophora,* does not occur in South Africa. It is unique and differs from the other five orders in having spinning glands which open on the anal segment by 1-3 pairs of spinnerets. In general it resembles the Juliformia more than the other orders, sharing such characters as the number and position of the defensive glands.

88

Park and in the northern and north-eastern provinces of southern Africa (Mozambique, Zululand, northern Transvaal and Zimbabwe), thus with a subtropical distribution. This handsome millipede has a striking livery of alternating black and yellow bands, football-jersey fashion, the head and neck segment jet black, the antennae and legs yellow (Figure 26).

The worm-like millipedes living in the strip of indigenous rain forest which fringes our coastline from the Cape to Zululand are rather different from those found inland on the high and sun-baked plateaux of the interior. They are in the first place smaller in size and include most of the brightly coloured kinds like *Chersastus*, while the inland forms are larger and almost entirely black or brown. In its forest home *Chersastus* is able to climb the branches of small trees, an adaptation brought about by its life in the forests and bush, out of which it seldom ventures.

## 3. *Distribution*

A few of our worm-like millipedes have a fairly wide distribution but most of them inhabit only a small area. One of those with a wide distribution is *Gymnostreptus pyrrocephalus;* it can easily be identified by its black body which contrasts strongly with the legs, head, and tip of the tail, these being a bright red; the distribution of this millipede extends all the way from Cape Town to Swaziland and perhaps the reason for its unusually large range is the fact that it is one of the few millipedes which breeds in cattle and horse droppings; the juvenile stages probably feed on the manure. Nor can the adults of this species be called dainty feeders for, unlike the great majority of millipedes, they show little discrimination in their choice of food. The reason may be that the population density of this species is greater than that of most other South African millipedes but it is certainly a scavenger and coprophage and has been seen attempting to eat pieces of old orange peel and scraps of dirty paper.

## 4. *Some enemies of Juliform millipedes*

It has been said in a previous chapter that millipedes have few enemies and that other animals do not as a rule prey on them; an exception must be made in the case of some of the stink-bugs or assassin-bugs (Reduviidae) which force their long needle-like beaks between the hard body rings of the millipedes and suck the unfortunate victims dry. Both large and small millipedes have been seen to be killed in this way by these bugs, which attack them in such large numbers that their bodies are almost hidden by clusters of the insects. The bugs are apparently not repelled by the evil smelling secretions of the millipede, perhaps on account of the fact that they themselves have the power to produce an offensive odour.

Miss Anna Rothmann formerly of the Albany Museum, Grahamstown, has written the following interesting account of her observations on these assassin-bugs attacking millipedes.

During the course of last March I happened on three occasions to find a millipede lying limp and immovable on the path in the early hours of the morning. On one occasion six young assassin-bugs (Reduviidae) were busily engaged in attacking it. They were so engrossed in the business on hand that they hardly moved when I picked them up, victim and all, and placed them in a small box for further observation. Within an hour they were so sated with feeding that their small bodies were almost round and the millipede was a mere empty shell. After a week spent in confinement they had again resumed their normal flattened appearance and all of them had moulted.

A little later I had the opportunity of observing the manner in which three fully grown bugs attacked their victim. They approached the millipede very cautiously and one after the other made a sudden leap upon it, piercing it quickly with their needle-like mouthparts, and then backing away. The millipede had obviously received a painful injury from these stabs which pierced its body between the chitinous rings. It reared upwards and then quickly threw itself into a spiral. Hardly had it recovered from this assault before the assassins were again upon it, piercing its body once more with their sharp beaks. These tactics were continued until the millipede was too far gone to hold off its attackers any longer and the assassin bugs were able to begin their meal at leisure.

The insects must be able to inject some sort of poison, or substance which has a paralysing effect, into the body of the victim, for otherwise the millipede would hardly have succumbed so quickly. It is interesting to note that these bugs, however hungry they are, will not touch the smaller red millipede, *Chersastus.* They attack only the larger blackish-brown kinds (*Doratogonus*).

These assassin bugs had also been seen previously attacking large spiro-streptid millipedes in the Kaokoveld of South West Africa where they had been photographed, and also in Natal. Miss Rothmann's Reduviid bugs were identified by the eminent entomologist, Dr A.Villiers of Paris, as belonging to two different species of the same sub-family, the adults being *Cleptria cinctiventris,* the larvae a species of *Glymmatophora;* the bugs of this sub-family appear to specialise in juliform millipedes on which to prey.

Another enemy of the worm-like millipedes in European countries is the common starling which destroys large numbers of them every year, but most birds seem to make a very definite distinction between insects and millipedes as an article of diet. None of our South African birds are known to make a habit of destroying millipedes by choice or of making them the chief item on their bill of fare (see page 82).

## 5. *Some parasites of juliform millipedes*

This account of parasitism in the Diplopoda is placed in the section dealing with the Juliformia since the comparatively few external parasites of millipedes are all found on members of this order.

The mite parasites of millipedes are of two kinds, the one consisting of mites which are permanent and specific for each species, the other of those which are sporadic and of a temporary nature, using the host merely as a vehicle for travelling from place to place.

Two small, round Gamasid mites are permanent guests of the large black spirostreptid *Doratogonus* in Natal, and quite obviously do their hosts no injury. They are commensals or mess-mates, living with them in a partnership beneficial to both parties; they do not have piercing or cutting mouth-parts and would be quite unable to prey upon the millipede by sucking its blood as the assassin bugs have been described as doing on a previous page. Two different but related mites live on each host, a larger, *Neomegistus* (Figure 27), and a smaller, *Paramegistus,* both having been first observed and meticulously described by the eminent Swedish entomologist Dr Ivar Trägardh. Nearly all the species of *Doratogonus* carry a number of both these mites which cling to the smooth body rings of the millipede with the aid of sucker-like pads which are present on all the legs except the first pair: they can run backwards, forwards and sideways with equal agility, using the natural groove formed between the insertion of the legs and the body wall as a long gang-way for running up and down the length of the millipede or when trying to evade capture with the forceps.

It was possible in 1936 to watch the performances of these lively parasites at fairly close quarters with binoculars in the thick bush on the Bluff at Durban. A puzzling question was how they could possibly find anything to eat in such an unpromising locale as a millipede's smooth body and if they did, what kind of food was it? An examination showed that they had weak brush-like mouth-parts only suited for sweeping up liquid food; they were evidently scavengers, pickers up of unconsidered trifles, and perhaps also playing a small role keeping the host free from particles of foreign matter.

Dr Trägardh when describing the biology of the mites in 1907 had been of the opinion that they fed upon the defensive secretions of the host. In my later paper on the habits of these two mites it was suggested that a substance repugnant to most animals and containing a large amount of hydrocyanic (prussic) acid could hardly been regarded as suitable food material. Furthermore it was difficult to persuade *Doratogonus* to secrete at all; it would tolerate some rough handling and even actual pressure before grudgingly producing a few small drops at the gland openings. Observations throughout the year brought to light the fact that the parasites disappeared during the cold months, probably overwintering in forest soil; in the summer they were back with the host again.

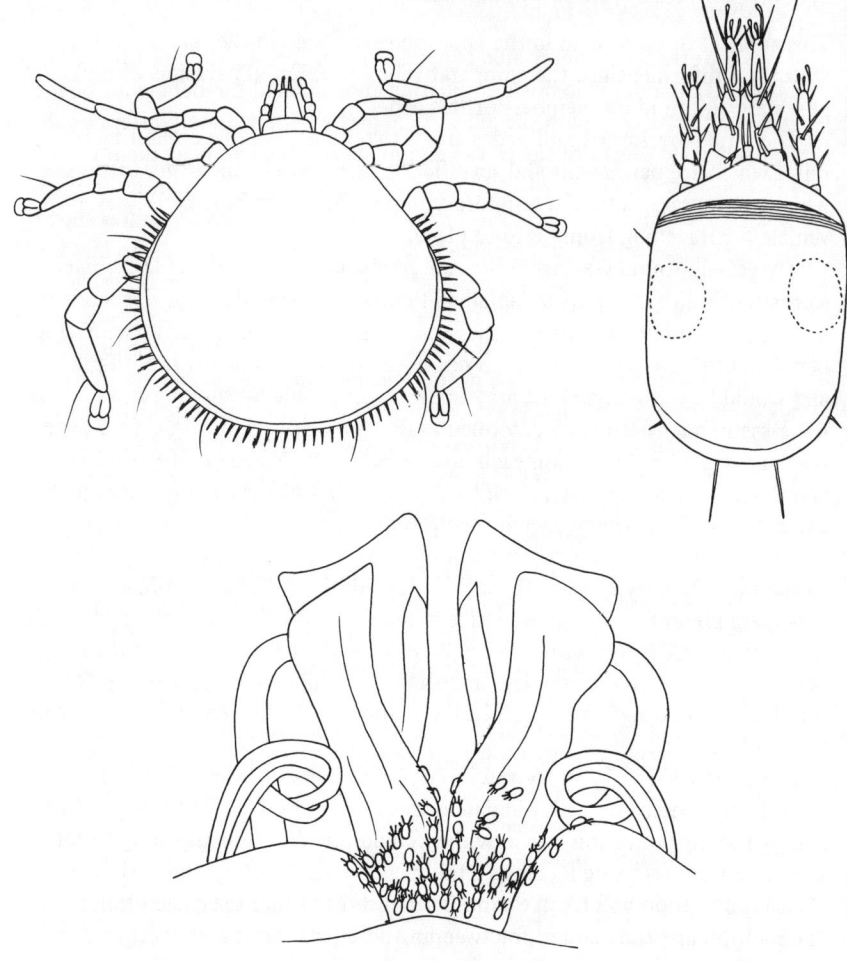

Figure 27. Mite parasites of the spirostreptid millipede, *Doratogonus setosus:* above left, the flattened gamasid mite, *Neomegistus;* above right, the deutonymph of the tyroglyphid mite *Caloglyphus;* below, the nymphs clustered on the sex organs of a male host.

For some reason unknown they were found in considerable numbers on the males, but very seldom on the females; the larger of the two parasites, *Neomegistus,* was entirely absent on 27 females examined while on the same number of males 91 were counted; the few *Paramegistus* mites, about one per female, were probably stragglers from the males during the sexual encounters of the millipedes.

Observations on the behaviour of the mites during the mating periods of *Doratogonus* during the summer months in Natal, can best be described by

92

the following excerpt from my paper.

> I was unfortunately never able to see the mites feeding though I repeatedly observed the millipede hosts copulating and watched the behaviour of the males subsequently with the aid of field glasses. Immediately after copulation the male carefully cleans the gonopods and the legs near them with his mouth, removing most of the traces of sperm in a few minutes (see also page 76 ). During this operation a number of the *Neomegistus* mites showed a definite interest in what was being done, standing on the collum and at the sides of the seventh segment, and passing repeatedly from one side to the other over the back of the host. One mite came very close during this process by approaching over the front of the head, eventually getting as far as the labrum; none of the mites remained quiet for very long, advancing as near to the gonopods as possible and then retreating quickly at the least sign of danger. The chief reason for their retreat seemed to be the continual movement of the legs and head of the millipede which appeared as if they might sweep them off; they carefully avoided all contact with the continually moving legs of the host.
>
> The mites may thus have to wait until the male has completed his toilet before they are able to take what is left of the sperm. It is improbable that the cleaning process is so thorough that some sperm fluid is not left in the crevices between the legs and sternum which would be sufficient to provide a meal for such small parasites. As soon as the attention of the host was engaged elsewhere they would be free to complete the scavenging process, an arrangement which would benefit both parties.
>
> This relationship between the millipede and its mites seems to be unique and it would also be reasonably probable that the liquid food of these mites is the seminal fluid of the millipede on which they live.

A rather different and, from the point of view of the millipede, far more deadly form of parasitism is practised by flies of the family Phoridae belonging to the genus *Megaselia*. They also attack millipedes such as *Doratogonus* but in another way. They lay their eggs on the body rings of the millipede, thereby according to Dr Brian Stuckenberg, to whom I am indebted for this information, 'causing great agitation' to the millipede. The eggs adhere to the side of the millipede and the larvae of the fly on hatching bore their way through the membranes connecting the body rings into the body of the millipede and feed on the soft inner parts of their host.

Temporary and adventitious parasitism in millipedes is illustrated by another quite different mite, *Caloglyphus*, belonging to the family Tyroglyphidae (Figure 27), the immature wandering nymph stage of which, the deutonymph, is found on various ground-living arthropods such as woodlice, insects and spiders, particularly those that live in the humus of the forest floor. They attach themselves in large numbers to the genitalia or sometimes the mouth

93

parts of the host animal by a cluster of powerful suckers on the underside of the body (Figure 27), these mites infest various species of *Doratogonus* and *Chersastus* as well as other Juliform millipedes. By using their hosts as carriers they can be dispersed far and wide thus contriving an extended distribution. They are minute and though occurring in vast numbers, appear to do the host no harm whatever; in any event they have at this nymphal stage not developed mouthparts for feeding.

## 6. *Introduced species of juliform millipedes*

One of our smaller worm-like millipedes, *Archiulus moreleti,* and an inhabitant of the Mediterranean region, has been introduced into South Africa by means of shipping, probably by way of Portugal. It has become very common in the greenhouses and gardens of Cape Town and also enters dwelling houses and flats in large numbers, climbing up the walls of rooms and making a minor nuisance of itself. Up to the present this is the commonest and most widespread member of the worm-like millipedes which is known to have been introduced into our country, and like many other foreign immigrants, such as the Argentine ant, has multiplied exceedingly in the south-west Cape and will probably spread further if steps are not taken to check it. All our other worm-like millipedes are native to this country.

## B. THE KEELED MILLIPEDES OR POLYDESMOIDEA

The members of this order are easily distinguished by their moderate size and the fact that they have comparatively few segments; the great majority of species have a fixed number of 19 or 20 (usually the latter). The body rings have a prominent keel or flange at each side which covers and conceals the legs (Figure 28a); very often they are provided with curious sculptures, rows of little tubercles or other ornamentation. Eyes are always absent throughout the entire order. The Polydesmoidea are of all the diplopod orders the richest in species and they display a great diversity of colour and form. Though the normal size is small to moderate (about 25 mm in length) the smallest species are only 4 mm in length while the largest forms, found in the tropics, are as much as 130 mm.

## 1. *General appearance and structure*

The head is rather small as compared with the Juliformia and is noteworthy in being without eyes; the antennae however have a peculiar finger-like organ, not found in any of the other orders, which may be sensitive to light and may serve as a direction finder. This organ is situated on the seventh segment and

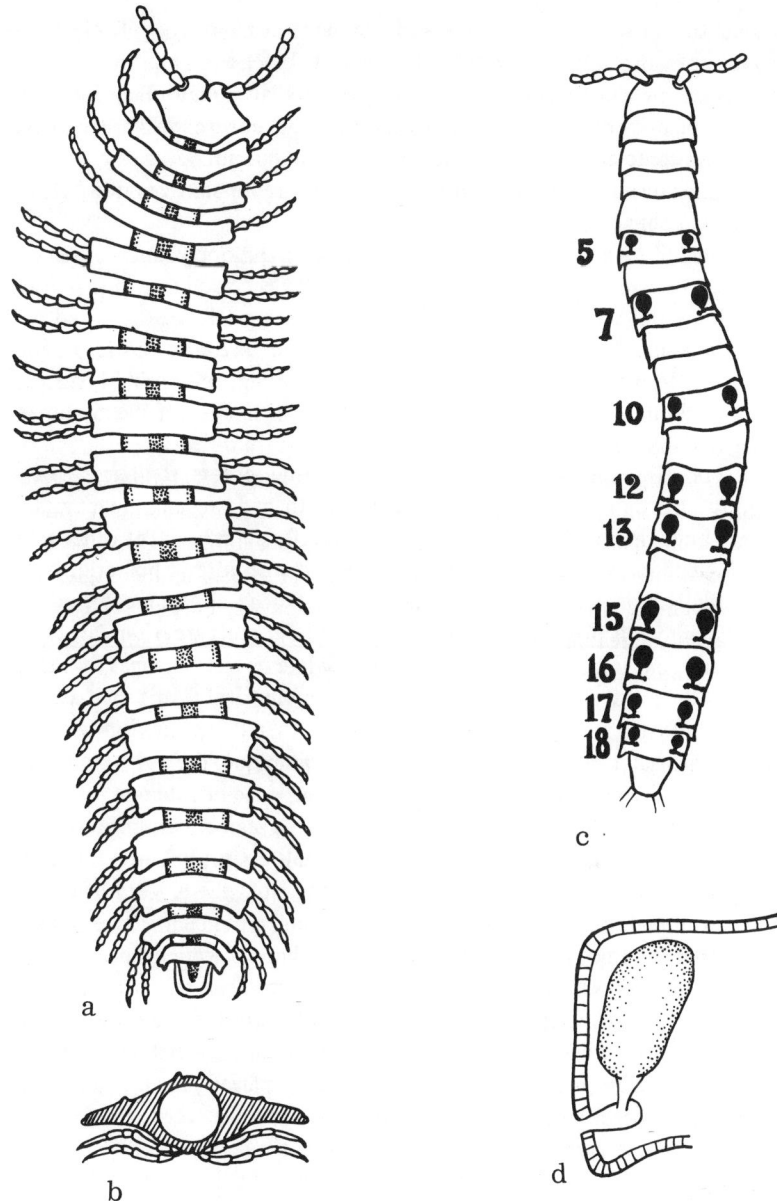

Figure 28. a, general view of a keeled millipede, *Polydesmus;* b, transverse section of a body ring; c, defensive glands in another keeled millipede (*Orthomorpha*) shown diagrammatically; d, a single gland enlarged.

has a large number of basal ganglia well supplied with nerves, while at its tip are a number of rod-like organs. At the apex of the vestigial eighth antennary segment are the usual four conspicuous sense papillae (Figure 21a). Situated near the apices of segments 5-7 are groups of rod-like spines which Verhoeff thinks are perceptors for apprehending changes in humidity. The organ of Tömösvary takes the form of a small pear-shaped depression situated below and a little behind the insertion of the antennae.

The mandibles are more complex than in other Diplopoda and consist of three main divisions, (1) a molar plate for grinding the food into fine particles, (2) the large and powerful teeth for biting and tearing off pieces of food material and cutting them into smaller fragments, and (3) the elastic lamellae, six in number, which are adapted for scraping, sucking, and probably also for straining the more liquid portions of the food. The cleaning of the antennae and legs are also performed by this section of the mandible.

The body is more strongly calcified than in other orders, the tergites of the body rings being fused with the sternites to form a single compact and strongly built ring, which in cross section is usually cylindrical, at other times more flattened in shape (Figures 28b, 29). The metazonite or posterior half of the ring is quite different from the anterior half in appearance on account of the lateral keels projecting from each side. These keels are often large and outstanding and may in some species have saw-like or fretted edges, in others they may be perfectly smooth; in many of the cylindrical forms such as *Gnomeskelus* (Figure 29), a huge genus of more than 80 species limited to indigenous coastal forest from the Cape to Mozambique, the keels are hardly visible or are represented only by a slightly raised ridge or tubercle. In general, however, these keels give the posterior part of each segment an impression of being much larger than the anterior half. The keels do not underlie those of the segment in front, being in fact well separated from them. Very often the whole upper surface of each posterior half or metazonite bears three transverse rows of nodules or tubercles, the size and shape of which vary among the different species. The whole tergite with its large keels and striking ornamentation can give members of this group a very characteristic appearance. In many species the tergites have regularly arranged hairs which may be spatulate, club-shaped or pointed; they are usually placed on the summits of the rounded tubercles which form transverse rows on the tergites.

The first body segment, the 'neck' or collum, is legless and without respiratory organs, the next three segments each have one pair of legs and a pair of spiracles, the remaining segments having two pairs of legs and two pairs of spiracles.

The openings of the spiracles are slit-like and protected by numerous minute chitinous spines. A closing mechanism is present. The tracheal chambers end in two horns and are provided with a muscular pumping mechanism for renewing the exhausted air; the tracheae arise for the most part from the

96

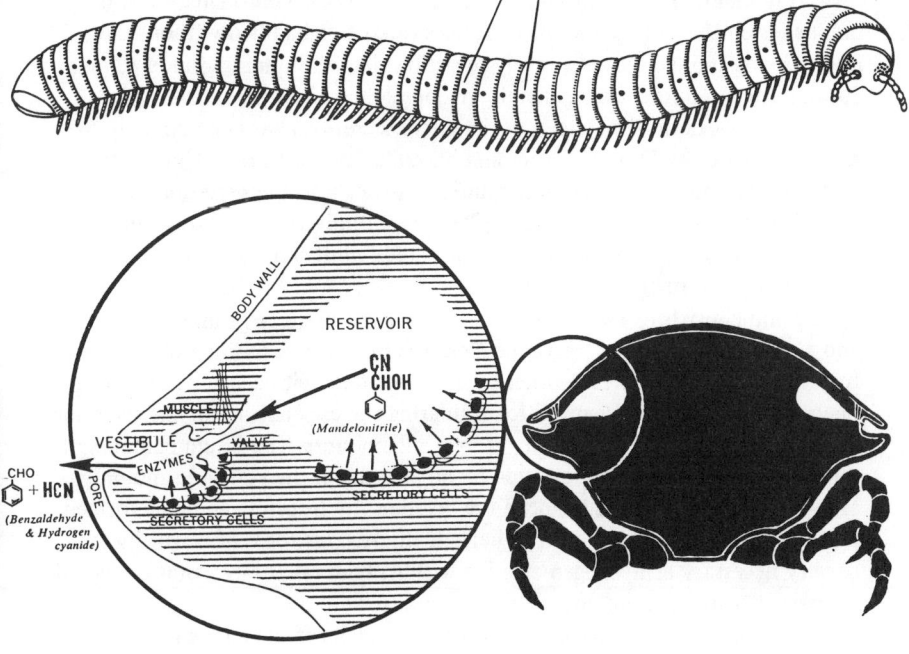

Figure 29. Above, position of the defensive gland openings (o) in a juliform millipede; below right, the defensive glands of a polydesmid millipede in transverse section; below left, the chemistry of the hydrocyanide producing mechanism shown diagrammatically (from T. & H. E. Eisner 1965).

basal half of the chamber and are similar in their general form, being long, narrow, and unbranched; they are reinforced with a regular spiral thickening of chitin.

The defensive glands are on the whole larger and better developed than in the worm-like millipedes; they are most frequently present on the following segments: 5, 7, 9, 10, 12, 13, 15-19, but are always absent from the sixth segment (Figures 28c, d). The substance secreted contains a large proportion of prussic acid and is thus extremely toxic. The expulsion of the secretion does not attract as much notice in this group as in the Juliformia, the ejection of the contents of the gland being a comparatively slow process. In the few cases in which it has been observed the secretion is colourless and has the usual odour of iodine. In South African forms the secretion must be very small in amount or difficult to evoke, as the appearance and smell of the odoriferous substance has hitherto only been noted in one species of *Platytarrus* (Figure 30), a polydesmid living at high altitudes in the Drakensberg mountains of

97

Natal which produces a copious acrid secretion of cream-yellow liquid with a smell resembling that produced by the European *Polydesmus.* In Juliform millipedes on the other hand many species cannot be handled without the fingers being stained by the actively secreting glands. Spinning glands, as in South African millipedes in general, are completely absent from the order.

The best work on the subject of defensive secretions has been done by the American biologist Thomas Eisner and his colleagues at Cornell University, who have carried out some illuminating researches into the secretions of Polydesmoidea as well as various other orders of millipedes, investigating the structure and working mechanism of the glands while determining the chemican composition of the secretions.

A gland consists of two compartments (Figure 29), a large inner 'reservoir' and a smaller vase-shaped 'vestibule' near the opening, each lined with secretory cells and joined to each other by a narrow duct with a special muscle, an opening and closing mechanism for regulating the expenditure of the secretion and when necessary only releasing it in small amounts. The liquid contents of the reservoir require to be activated by an enzyme in the vestibule and only after the two have been mixed can the resultant hydrogen cyanide be formed together with a certain amount of benzaldehyde. This form of control would be very necessary as hydrogen cyanide with its high rate of evaporation would be very difficult to store under natural conditions.

The secretion is very effective against the attacks of the ants which occur in large numbers in the woodland environment in which the millipedes live. It was also found to repel toads, carabid bettles, some birds, armadillos, opossums and skunks.

Other orders of millipedes investigated by the researchers at Cornell University were the Colobognatha (the American *Polyzonium*) which produces a secretion with the odour of camphor, and the Nematophora, absent from the South African fauna, which secretes a substance smelling like phenol or creosote. Various spirostreptid millipedes, including a species of *Doratogonus* from Lesotho, produced quinones.

There is only one pair of gonopods in the male instead of the usual two pairs of other Diplopoda; as in other millipedes it is situated on the 7th segment, and is represented by the modified first pair of legs, the second pair on this segment being normal walking legs. The gonopods are conspicuous organs protruding through a rounded aperture with a rimlike edge in the anterior half of the seventh sternite (Figure 23). The vulva and the paired openings of the male reproductive organs are situated behind the second pair of legs, that is on the third body segment. Secondary sexual characters are not as well marked as in the Juliformia but the bodies of the males are distinctly more slender while the legs are usually longer and stouter than those of the females. In addition the undersurfaces of the tarsi, and sometimes also of the tibiae, of the first three legs of the male carry a brush of modified spatulate

hairs. The anterior tarsi in the largest South African family, the Sphaerotricho-
pidae, are provided with numerous small round tubercles. The tarsal brushes
of the male are used for cleaning the antennae (Figure 23).


## 2. *Reproduction*

Pairing and egg-laying take place in the spring, the eggs being laid about a
month after copula. They are minute and round and in some species of polydesmids are deposited in a special nest by the female. It is made of moist
earth voided by the female and moulded into shape by the anal flaps, a more
or less circular membrane which can be extruded from the anus, and the last
pair of legs. A fairly plentiful secretion from the anus is used to cement the
building materials. In the European species, *Polydesmus complanatus,* the
eggs are laid two at a time and during the egg-laying the female frequently
pauses to build up the sides of the cup-like nest a little further. The rising
walls curve over towards the centre so that the nest assumes a dome-like shape.
When the dome is completed a ventilating chimney is added in the middle and
the millipede disguises the nest with fragments of earth. The whole structure
may take from 6 to 24 hours to make, and its materials are used by the newly
hatched larvae as food. The number of eggs in the nests seems to vary very
considerably but a large number are usually found. In a nest of *Gnomeskelus
silvaticus,* observed by the writer at Knysna, 370 eggs were counted, each a
little smaller than a pin's head. In this nest the opening at the top was large
and the nest uncamouflaged. The female was curled round the opening at the
top of the structure blocking it completely with her body. An American
investigator H.Miley counted 526 and 586 eggs in the case of a North American polydesmid, *Euryurus erythropygus.* No elaborate nest was made by this
species, the eggs being deposited in small cavities made a little below the surface of the soil by the female.
    The young millipedes hatch out 12-15 days after the eggs are laid and the
newly emerged larva has seven body segments with three pairs of legs, one on
each of segments two to four. No food is taken by the larva for the first few
days. The second stage has nine body segments with six pairs of legs, the
third 12 segments with 11 pairs of legs, and the fourth 15 segments with 16-
17 pairs; the seventh or pre-adult stage has 19 segments and 28-29 pairs of
legs and from this the adult emerges with its 20 body segments and 30-31
pairs of legs. Between each successive stage there is a moult, but ecdysis is
also continued after the adult stage has been reached and apparently goes on
throughout the life of the millipede. The larvae make a mud nest in which to
undergo the moult resembling that made by the adult female for egg-laying,
but smaller. While however the brood chamber is constructed from the outside by the adult female, the young larvae make their moulting chambers
from the inside, the final act being to close the aperture at the top of the nest

**99**

so as to secure it against predators at this critical time. The observations of Miley confirm the statements of previous authors that the larvae make moulting nests during all of their seven developmental stages. The first nests must therefore be very small, increasing in size as the larva grows, while the last preadult moulting nest is only a little different in size and appearance from the brood chamber of the adult female.

The newly hatched larva requires 11-12 months to reach the adult condition and the adults themselves live for 2½ to 3 years and reproduce more than once during that time; thus in nature young at various stages are almost always found together with the adults. During the breeding season copulations by the same individuals are frequent, and may last hours or even days; the mating pairs can be quite roughly handled without fear of separating them.

### 3. Moulting and other habits

Moulting takes place among the adults in a specially built nest which is similar to that made for depositing the eggs except that there is no opening at the top; in some cases however moulting may be carried out in a suitable cavity in the ground or in rotting wood. The animal goes into a quiescent state as the exoskeleton becomes softened by secretions of the skin glands which assist in lifting the old skin from the underlying new cuticule; the old skin eventually splits between the head and collum. The linings of the fore and hind gut are shed as are also those of the mouth, tracheae, and gland openings. The new soft skeleton soon hardens and only then does the animal resume its activity and again take food.

The movements of the Polydesmoidea are slow and clumsy as compared with many of the juliform millipedes. The keel-like outgrowths of the body segments set certain limits to the flexibility of the animals and they are unable to make sinuous movements or to enroll in a complete spiral. When disturbed they either remain rigid or curl themselves into a rough spiral, their main defences consisting of the odoriferous glands and their concealed manner of life. They are light avoiding and retreat into crevices in wood or under bark during the day, coming out at night to seek their food and to mate. The food is entirely plant material, consisting of decaying wood and leaves, of which there is an abundance in the situations where they live. Their habit of burrowing into and feeding upon woody tissues plays a certain part in breaking up soil and decaying vegetation, and thus assisting in the formation of humus.

The often observed toilet habit, or cleaning and polishing of the antennae and legs, is carried out by certain structures of the mandibles with the assistance of secretions from salivary and other glands situated in the head. Members of the group seem to be free from external parasites, though gregarines and nematode worms have been found in the intestine.

The Polydesmoidea are the only order of millipedes in which luminescence occurs. Though unknown in our South African species, certain polydesmids such as *Amblocheir sequoia* in South California are continuously luminescent, the whole body glowing with a greenish light.

## 4. *Distribution and numbers*

The nature and abundance of the food on which they live makes it unnecessary for polydesmids to move very far from the localities of their birth so that the species are fairly localized. In general members of the group are found everywhere throughout the world with the exception of the polar caps. The tropical regions have a greater abundance of species than the temperate ones. Many species have become widely distributed by artificial means and have been introduced into various countries in soil, hothouse plants, and timber. In South Africa such a species is *Orthomorpha gracilis* which has become firmly established as a minor pest in gardens and greenhouses at Cape Town and other coastal towns. This millipede has also reached some of the inland centres of the Republic as it has been found many hundreds of feet underground in timber props of mines at Kimberley and Johannesburg. One small, entirely pale species, *Phygoxerotes nodulosus* Verhoeff, is undoubtedly a myrmecophil and is found in various parts of South Africa in the nest of the black ant *Camponotus thraso* Forel. Silvestri has described a number of polydesmids living in the nests of termites in South America. Another new form, *Attemsodesmus*, found in ants' nests at Port Edward, Natal, is only 3 mm in length.

All the South African Polydesmoidea are small to moderate in size, without strongly contrasting colours, though several have a pretty pattern of subdued tones. They are all confined to the forest floor of indigenous bush or to small localities where the soil is rich in humus. Only one of the larger forms, *Ulodesmus*, damages cultivated plants, but even in this case the attacks are light and infrequent. About 150 species accommodated in four families are known from South Africa, 112 from the Natal-Zululand region alone. In general the arid inland plateaux have a poor polydesmid fauna, and from South West Africa, the western parts of the Transvaal, and the Orange Free State, extremely few species are known.

## C. THE PSELAPHOGNATHA OR PENICILLATA
Dwarf or 'pin-cushion' millipedes

The Pselaphognatha are the most primitively organised of all the groups which make up the heterogeneous class of Diplopoda. In the first place they differ from nearly all other diplopods in their uniformly minute size which is seldom more than 4 mm in total length. In the second place there is no vestige of a calcareous skeleton, the skin being soft and delicate with a few chitinous thickenings for the better support of the body. In the third place there are numerous peculiar hairs or trichomes which are arranged in groups and tufts like small pincushions or brushes (Figure 31a).

This dwarf millipede though so different in outward appearance has rightly been regarded as a diplopod and this is proved by the fact that the body segments are double or diplosomites, which is the hall-mark of the class: furthermore the short antennae have the typical number of eight segments. Thus, though a divergent one, *Polyxenus,* the dwarf or pincushion millipede, is a true diplopod.

### 1. *General appearance and structure*

The head, which is distinct from the body, is moveable and can be turned a little. It consists of an anterior and posterior section divided by a deep cleft. The anterior part includes the mouthparts, the posterior the antennae, eyes, and trichobothria (Figure 31e), long delicate movable hairs which are sensitive to air currents and vibrations. The number of the small ocelli is variable (five to ten or more) and in some species they are absent altogether. They are situated on the lateral margins of the posterior half of the head, and just below them but a little nearer to the middle is a group of three sensory hairs, the trichobothria (Figure 31e), always arranged in the form of a triangle; they are only present in the Pselaphognatha, being unknown in any of the other diplopod groups, and are considered to be organs of hearing. Below each antenna and a little to the side is the minute organ of Tömösvary, which is however present in only one genus, *Lophoproctus,* not found in South Africa.

The minute mouthparts have a biting rather than a sucking function. Some of the structures found in other Diplopoda are either undeveloped or missing; the single-jointed mandibles are well developed and though they lack large teeth there are plates with numerous fine denticles arranged like the teeth of a file which can deal with minute particles of the plant material which probably constitutes the food of the animal.

The body consists of 11-13 segments, the first being legless, the next three having one pair and the remainder two pairs of legs, with the exception of the anal segment which is also legless, the whole arrangement typical of

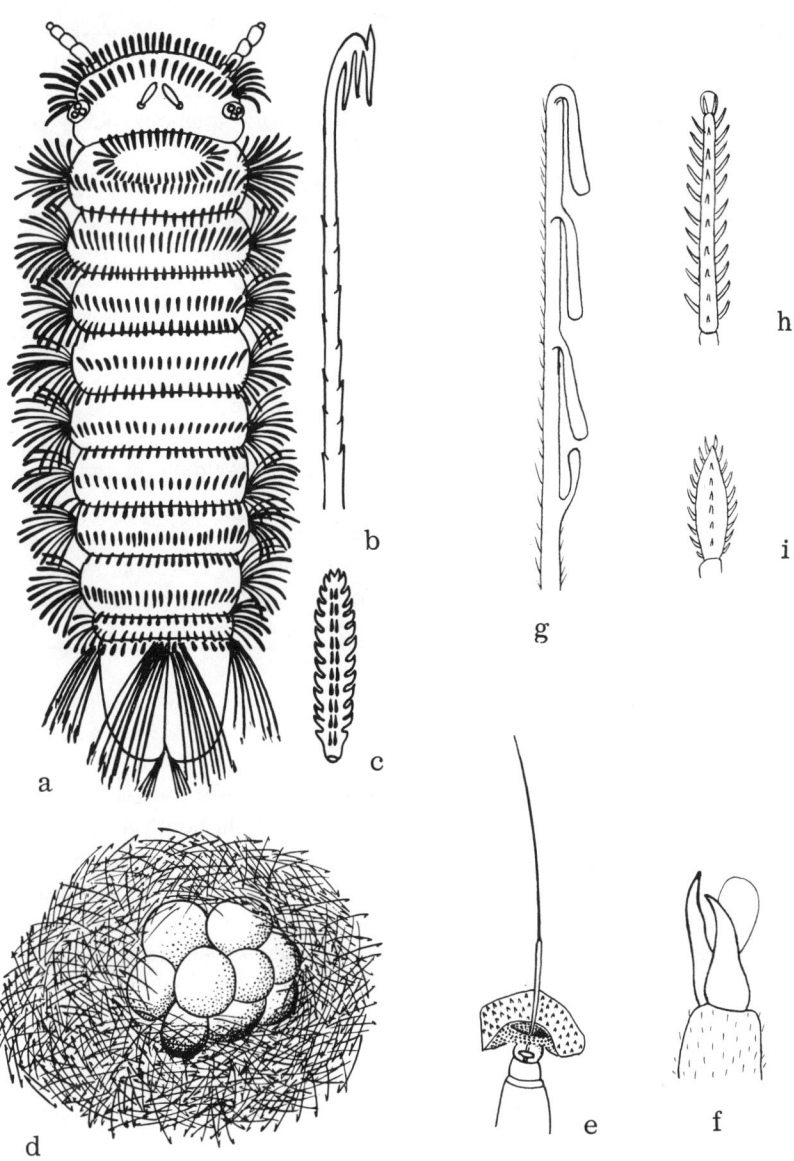

Figure 31. a, generalised figure of a pincushion millipede (Pselaphognatha); b, along hair from the tail segment; c, a body hair; d, the 'nest' with eggs; e, one of the three peculiar trichobothria at the side of the head enlarged; f, claws of a leg with adhesive lobe (all from the European form Polyxenus); g, a long hair from the tail segment, and h, i, body hairs of a species of *Monographis* from Port Alfred, East Cape.

103

the millipede class. The legs end in a thick-set claw provided with a large adhesive pad-like organ which enables the animal to climb smooth and almost vertical surfaces, or quite often to walk upside down (Figure 31f).

The clothing of the body is peculiar and unusual in Diplopoda, being composed of numerous specialized hairs arranged in tufts or bundles, the hairs radiating outwards like the pins of a pin-cushion.

These hairs or trichomes which by overhanging the short legs give them complete protection, are of various shapes and patterns and are hollow and partly filled with air. They fall off very easily, especially the longer trichomes which in some species are provided with retroverted barbs at their apices giving them the appearance of a shepherd's crook (Figures 31b, g); those situated at the posterior end of the body are of this type and many of them are roughly woven together by the female to form an enclosure or rough nest for the eggs; they may also serve as defensive organs. Shorter hairs and sometimes scales may be found on other parts of the body, the patterns and shapes varying from species to species (Figures 31b, h, i).

The spiracles are attached to the chitinous thickenings of the coxae and are found in all segments. The tracheae are relatively wide and soon divide into a comparatively smaller number of branches provided with a spiral thickening. In general they resemble those of the Oniscomorpha.

## 2. Development and moulting

There are two penes but no gonopods in the male and copulation is not known. The openings of the reproductive organs are situated between the second pair of legs. Nothing is known about the breeding habits of our South African dwarf millipedes but according to the observations of C.Hector on the New Zealand *Polyxenus* the eggs are laid in the early spring in small compact groups of nests, each nest containing 4-8 eggs of relatively enormous size (0.3-0.4 mm). The nests are provided with a covering of roughly interwoven hairs from the tufts at the end of the animal itself and cemented with a sticky liquid secreted by glands which form parts of the sex organs (Figure 31d).

The young leave the egg with three pairs of legs, and stages with 4, 5, 6, 8, 10, 12 and 13 pairs follow in the case of *Polyxenus lagurus.* There is a moult between each successive addition of leg pairs. In this species, *Polyxenus lagurus,* no males have been found among many hundreds of specimens collected in certain parts of Europe and it is considered to be parthenogenetic in these regions.

In view of the absence of a calcareous skeleton moulting is rather different from that of other diplopod groups. The numerous hairs or trichomes exercise a strong mechanical pressure under the old skin and in this respect the process resembles that of insects rather than Myriapoda.

## 3. Habits and distribution

Little is known of the habits of these dwarfed millipedes. They live singly or in small groups in semi-decayed or damp wood, under bark or stones and are essentially forms which require a constantly humid environment. There are no defensive glands or spinning glands.

The European form, *Polyxenus*, with aid of the sucker-like pads on the feet is able to live, sometimes in colonies, upside-down on glass-smooth ceilings, such as are provided by the undersides of large flat rocks, and the cast skins left on such rocky surfaces provide evidence that they are able to pass much time in such confined spaces which are too smooth to allow predators to follow them. They can run out of their retreats for the algal food they require and return to their ceilings. In spite of its small size *Polyxenus* can with its flexible body walk surprisingly fast, looking like a pin-cushion on the march. When touched or irritated it often makes little short jerky leaps. The cuticle and the curious spines which cover it are both hydrofuge so that the animal is unwettable.

The easily detachable trichomes with their retroverted hooks may perhaps be regarded as their only means of defence if, as is conjectured, they contain some irritant poison such as is found in the bristly urticating hairs of certain lepidopterous caterpillars. For the rest their small size, natural agility and extremely cryptic habitat afford them sufficient protection to insure their continued existence.

The Pselaphognatha like other primitive arthropods form a very compact group, the genera differing but little from each other. About 41 species divided into 13 genera are found in various parts of the world. They are definitely cosmopolitan in distribution. Four genera and six species are known from South Africa and of these *Phryssonotus* is by far the most widely distributed, being found in the Cape, Transvaal, Mozambique, and Natal. Three of the genera are from Natal, the fourth, *Chilexenus,* is confined to semi-arid regions such as the Kalahari, north-west Cape and south-west Africa.

A species of *Chilexenus* has been found at Port Alfred by Dr P.Hulley living under smooth flakes of shale from the built-up walls confining the east and west banks of the Kowie river near its mouth, and only a metre or so from the water; it is thus probably able to tolerate a certain amount of salinity. On account of their small size and obscure living quarters the Pselaphognatha may have previously been considered to be uncommon myriapods; it seems more likely that they will be found to have a wide, perhaps ubiquitous distribution in South Africa.

## D. THE ONISCOMORPHA OR PILL-MILLIPEDES

The pill-millipedes belong to a group of Diplopoda so different in their appearance from what is usually understood by a millipede that quite a number of people would be unable to tell at a glance to what division of the animal kingdom they belong. One expects a millipede at least to be long and worm-like with plenty of legs, but the body of the pill-millipede is round and clumsy, and the legs, if they are seen at all, are inconspicuous and few in number. The pill-millipede, or to give it its scientific name, *Sphaerotherium*, meaning 'round-animal', is a good example of a creature in which every part of the body has become moulded and shaped for a single purpose. In the case of *Sphaerotherium* this adaptation is directed towards ensuring the safety of the animal by enabling it to roll up into a round ball at a moment's notice like a hedgehog, and a fairly large part of its time must be spent in this posture, all its vulnerable parts, the mouth, legs and soft under surface, being safely hidden beneath the covering of its hard armour-like exterior. This clumsy and peculiar millipede has, like one of the ancient armour-plated reptiles, sacrificed speed and grace for the tactics of passive defence (Figure 33).

### 1. *Enrollment or conglobation*

The body is covered by about a dozen curved shields or tergites which are immensely strong and so designed that they fit perfectly one behind the other, overlapping from in front backwards. The pill-millipede is enabled to achieve this tight method of spiralling by means of a number of adaptations; the body is a half cylinder in cross section and is so flexible that it can be bent strongly downwards to obtain tight enrollment; the imbricating tergites are strongly narrowed at the sides and fitted with a kind of flange which allows them a good deal of play as they ride over the neighbouring tergite; the head and first segment are small and can be tucked well in and partly covered by the large and powerful second segment, the head shield or *brustschild*. Finally these are again partly covered and protected by the last segment, the hood-like tail shield or pygidium which is very large and close fitting (Figure 32). The legs are greatly flattened from base to apex looking like a series of paper knives lying side by side, so that they take up very little space. Once enrolled there is no chink or crevice in the armour and the pill-millipede can defy the attacks of nearly all carnivorous animals or such small tormentors as ants.

So tightly does the armour fit that even water cannot enter when the animal is submerged. In addition to being able to defy predators the pill-millipede, when enrolled, uses up less vital energy and requires less air and food, while life goes on at a slower tempo. So strong are these protecting body tergites that, in the larger individuals at any rate, the strongest pressure of the hand is insufficient to crush them; most pill-millipedes are extremely smooth, shiny and polished so that they are almost slippery to hold; the whole appearance is one of extreme cleanliness.

The enrolling habit has an ancient history, having perhaps been invented by the marine trilobites, primordial arthropods which lived in the Cambrian seas 300 million years ago. There is an extraordinarily close resemblance in outward bodily form and appearance between the trilobite *Phacops* from the middle Devonian period and the modern South African pill-millipede *Sphaerotherium*. The stratagem has been made full use of by large land animals, mammals and reptiles, as well as by small invertebrates such as cockroaches and woodlice (Figure 32). Many minute arthropods living in the humus of the forest floor practise it, among them a tiny mite *Phthiriacarus* looking like a round shiny seed and a pigmy scarab beetle *Philharmostes*, both about 2 mm in length when enrolled. A small number of keeled millipedes, Polydesmoidea, also do it. All these have developed techniques for total conglobation in which the vulnerable structures are covered and concealed, affording them complete protection.

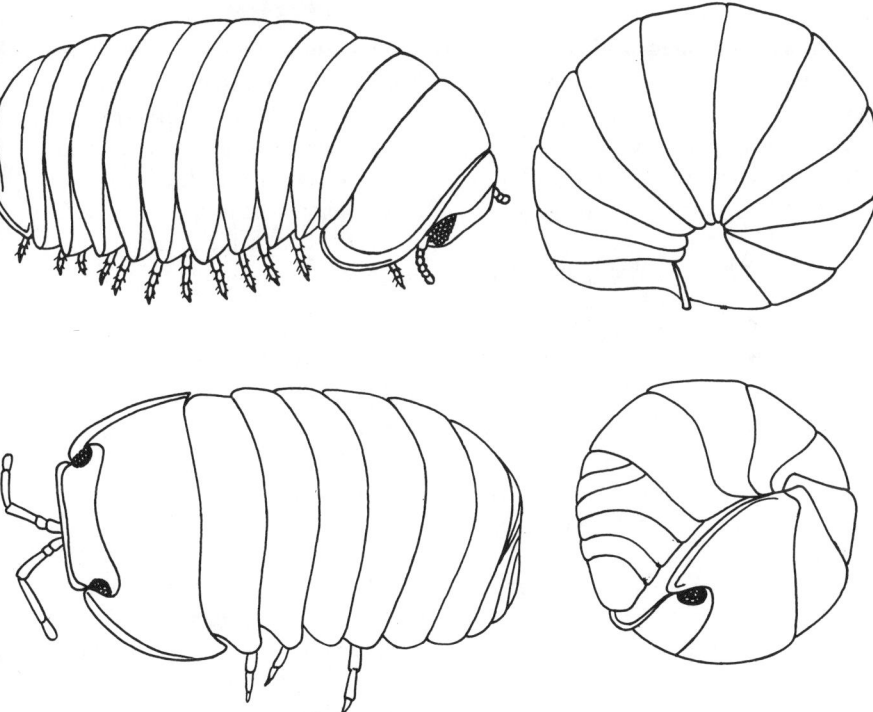

Figure 32. Above, a pill-millipede, *Sphaerotherium*, and below, a woodlouse, *Cubaris*, in extended and (right) completely enrolled attitudes.

## 2. *General appearance and structure* (Figures 32, 33)

The short body seen from above is oval, and covered by 11-13 tergites; the 21 pairs of legs and the antennae except for their apices remain hidden by the large covering tergites. The short, wide head and small collum are followed by the very large second tergite or shield (*Brustschild*) which has a distinct raised rim like the brim of a hat; when enrolled the last segment, the large pygidium, engages and fits tightly against it. The surface of the tergites are as a rule very smooth and shiny, but in some species may be pitted with innumerable small pores or carry a thick coat of short, fur-like hairs. The very large pygidium has a distinctive shape and can best be described as the half of a bell which has been equally divided down the middle.

The legs are weak and unfitted for vigorous movement — being strongly flattened and only loosely connected to the ventral sternites; the last two pairs in the male have been modified into clasping organs or telepods for holding the female in the act of copulation; the anterior of these is thus pincer-like and bears at the side a series of strong ridges forming a stridulating or sound producing organ of a type which is so widespread among the arthropoda (Figure 34b); it rasps against a patch of strong dark coloured tubercles on the inner surface of the pygidium.

The slit-like spiracles of the respiratory system lead into a large tracheal sac which can be roughly compared in shape to a wide based triangle which gives off the two main tracheal trunks (Figure 34a). The whole respiratory system is generously developed, the large and conspicuous main trunks branching into smaller ones which ramify over the whole body. The tracheal sacs, which are extremely large, play the part of a pumping mechanism with the aid of attached muscles; the repeated contractions of the body which accompany the enrolling habit probably stimulate the pumping of fresh air and its circulation in the tracheae.

The vulva of the female is to be seen on the coxae of the second pair of legs; it can be readily distinguished as two triangular, strongly chitinised plates or valves lying side by side, not to be found on any of the other legs. The openings of the male sex organs are also on the second pair of legs but are more difficult to distinguish. There is a lighter spot near the inner apices of the coxae of these legs which is easy to discriminate from the surrounding dark chitin; situated in the middle of this spot is the small transverse slit-like opening of the *vas deferens*.

The males of *Sphaerotherium* are always a little smaller than the females; for the rest the secondary sex characters are for the most part confined to the large tail shield or pygidium which usually differs from that of the female with its posterior surface often strongly depressed in the middle like a saddle, and with a well turned up rim. There are also other structures such as hairy pads, ridges or granular patches, which if present at all in the female are much less pronounced.

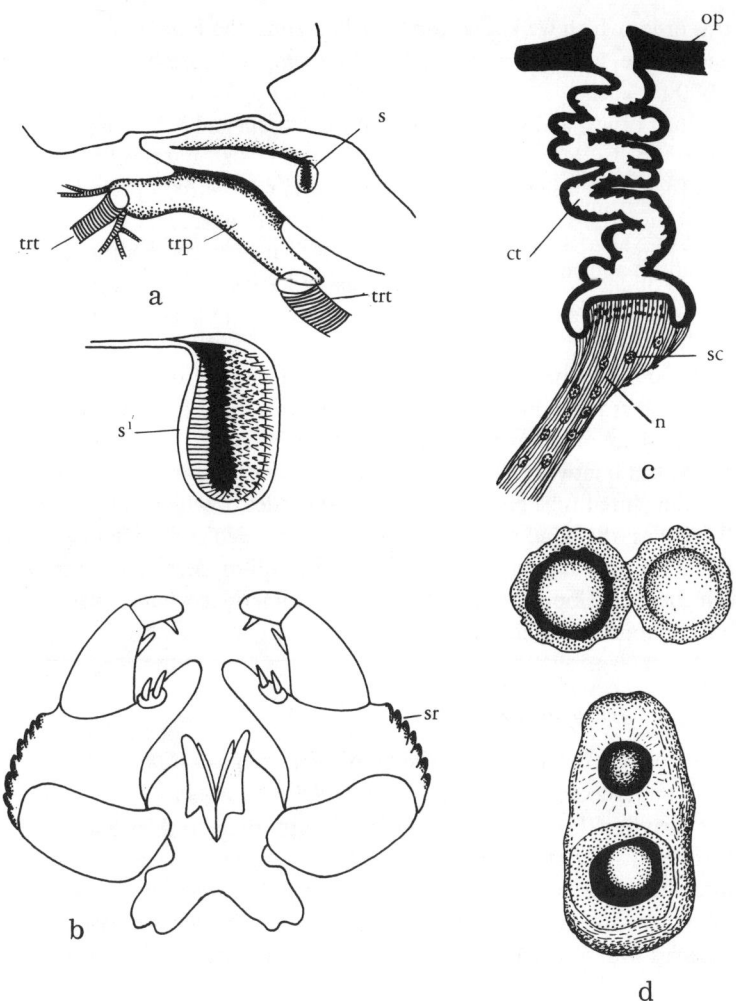

Figure 34. Structures of Oniscomorph millipedes. a, respiratory system with tracheal pocket (*trp*), tracheae (*trt*) and spiracle opening (*s*). b, the last, sexually modified pair of legs, the telepods, with stridulatory ridges (*sr*); c, organ of Tömösvary, an auditory sense organ, showing the convoluted interior (*ct*), sense cells (*sc*) and nerve fibres (*n*); d, the double chambered earthen egg capsule; and d[1], a smaller capsule containing a single egg.

## 3. *The sense organs*

The large group of eyes at the side of the head and below the antennae resembles that of the worm-like millipedes in consisting of a compact cluster of ocelli (Figure 20c), (about 60 in *Sphaerotherium giganteum*) differing somewhat in size and arranged in several rows. In some of the ancient milli-

109

pedes of this order which we know from fossil remains, the number of ocelli was extremely large, in some species such as *Glomeropsis crassa* being as many as 1 000 for each eye. Some of the ocelli of our *Sphaerotherium* are much larger than others and separated a little from them and it is thought by the German authority Verhoeff that these variations of size and position enabled it to focus objects at different distances and to receive impressions from different directions. The antennae are short and thick and have the usual olfactory organ at their apices composed of four little cones although some species of *Sphaerotherium* are exceptional in having a much larger number (18-20), each being supplied with a small branch from the antennary nerve.

*Sphaerotherium* has one sense organ not found in some of the other orders of millipedes such as the Juliformia and Colobognatha. What is generally agreed to be an organ of hearing, the organ of Tömösvary (Figure 34c), is situated on the head between the antennae and the eyes; its round opening leads into a convoluted tube about 200 $\mu$ long, the sides of which are covered with small spicules; the blind basal end is slightly expanded and its floor lined with sensory cells supplied by branches of an auditory nerve. The organ reaches its highest development in the European Glomeridae where it has an extremely large horse-shoe shaped opening.

### 4. *Mating and transfer of the sperm*

A study of the pairing of the sexes in the South African genus *Sphaerotherium* has been undertaken only once, that by Dr Ulrich Haacker of Darmstadt during the late sixties. The termination of his contributions to this complex problem through his untimely death at an early age, is a severe loss to biology and to South African zoology. The description which follows is based almost entirely on his account of the progressive mating stages in *Sphaerotherium dorsale,* a species widespread in the forests of the eastern half of southern Africa.

Pairing takes place at night and the mating partners are easily disturbed by bright light or other stimuli. In the presence of a female, the male approaches her, walking backwards; contact is made and he then levers her up with his shovel or scoop-like pygidium, rolling her over or pushing her along the ground; at the same time the posterior telepod with powerful ridges on the penultimate segment is extended backwards and begins to rasp intermittently against the tubercles on the inner side of the pygigium, giving rise to the shrill sounds which have been described by various auditors as squeaks, the vocal sounds of mating frogs, or even the screeches of a small mechanised toy.

In *Sphaerotherium dorsale* the 'squeaks' or impulses were emitted in a series or group of two to four with a pause between each impulse of 1-4 seconds; the call was repeated and another series of impulses commenced after a delay of about 40 seconds. The biological significance of this stridulation call seems to be that it represents a signal to the enrolled female that the waiting male, ready for the mating act, is one of her own species. The sounds form a pattern which when more species of pill-millipedes are investigated, will probably be found to be a code which is specific and only has significance for one particular species.

The female must have an auditory sense enabling her to receive and respond to the call (the organ of Tömösvary) for she unrolls and the actual mating can proceed; the male walks backwards passing his pygidium beneath the forepart of the female and with the widely opened pincer-like claw of the posterior telepod, firmly grasps her vulvae from beneath; the two fall upon their sides. The male bends his body further ventrally so that partners lie front to front but with their heads pointing in opposite directions; the sex openings are then only about 2 cm apart.

While the female lies motionless the male begins to move his legs with a movement differing only from that of walking in that the tarsi of each leg pair bend inwards and touch each other in the middle; this seems to facilitate the production of sperm and after 1-2 minutes a round spermatophore appears, 0.8-1.2 mm in diameter, at the sex opening of the male. The tailless non-motile spermatozoa are embedded in a sticky ground substance which forms an enclosing membrane round the sperms. The spermatophore is then grasped by the tarsi during the continuous movement of the legs which pass it backwards with an action like a conveyor belt and attach it to the vulva of the female, the whole exercise of transfer lasting about 10 seconds. The leg movements of the male then cease and both partners lie quietly for 2-5 minutes on their sides, after which the male releases the vulva and moves away, 4-7 minutes after pairing had begun. No cleaning of the telepods, as is customary in juliform millipedes, was observed.

As soon as contact is broken the female curves the fore part of her body downwards, bringing the head near to the vulva holding the spermatophore and, using the anterior legs and mandibles, pushes it into her mouth. It is not certain whether this is for the purpose of keeping the spermatophore clean or for eating the remains after the spermatozoa have left it. As a possible explanation Dr Haacker suggested that the mouth cavity, as in the Symphyla, was being used as a temporary receptacle (*receptaculum seminis*) for holding the spermatozoa. In European Oniscomorpha pairing is different in a number of respects and in *Glomeris* it ends with the male bringing a small ball of soil or faecal matter and after rolling it towards the female, applying it to her vulva.

5. *Egg-laying and development*

Little is known of the further development in *Sphaerotherium* of these processes and Dr Haacker in his observations recorded above says that no egg-laying had taken place in his specimens of *S.dorsale* which were kept in captivity for six months after pairing. The general outlines of the procedure must however be similar to those of the Oniscomorpha of European countries.

The laying of the eggs in the European *Glomeris* takes place in the spring. The female lies on her back or side and with the last pair of legs moulds a little egg chamber or capsule from pellets of faeces which are voided from the hind end of the body. About 18 of these capsules are fashioned by the female, each of them containing a single egg; when she has finished making an egg case and laid the egg in it, she closes it, pressing it into suitable shape with the posterior legs and anal valves; she then drops it, taking no further interest in its fate. The egg case usually has thick strong walls and often contains two compartments (Figure 34d), in one of which the egg is placed, the other being left empty; the egg chamber of *Glomeris* is round or oval and 3-5 mm in length.

When the egg hatches out after about 30 days the minute first larva spends some time within the egg case as the chamber is large enough to allow it to move about freely; some authors have stated that it uses the walls of its own prison as food and so manages to eat its way out at the right time; Verhoeff however is not of this opinion and it is perhaps more likely that a section of the nest is thin-walled and can be easily broken by the emerging larva. Early development takes place within the egg membrane where a stage with three pairs of legs and two to five very rudimentary bud-like ones, is reached. On breaking through the egg membrane it presumably can move freely within the egg chamber and it then has three pairs of segmented legs and five pairs of rudimentary ones which soon develop into segmented ones. When the larva eventually leaves the chamber it does so with seven body segments and eight pairs of fully formed segmented legs; stages with 9, 10 and 11 body segments follow and from the last of those the adult pill-millipede emerges.

The European representatives as far as is known live two to three years and reproduce several times; during that period moulting is carried out in a roughly made nest. Some of the large South African forms, such as *Sphaerotherium giganteum,* probably have a longer life span judging by the scarred and often battered appearance of the external skeleton in some of the older specimens.

6. *Habits*

The pill-millipedes are clumsily built animals and are the slowest moving and least agile of the various orders of millipedes. As their appearance indicates they are sedentary in habit, spending most of the daylight hours in a state of complete quiescence, unrolling again at night to move about and feed; if met with during the daytime it will be in damp shady places and if disturbed in the course of walking a millipede quickly reacts by enrolling. They are poor

112

burrowers but with the aid of the large and powerful head shield they can brush aside leaves and woody debris in their progress through loose woodland soil and can make shallow excavations which are however sufficient to cover and conceal them; more often they will be found lying enrolled beneath a large decaying log.

Although the enrolling action is quick and immediate, walking is clumsy and slow; when ascending a rough slope littered with obstacles, they sometimes lose their balance, tumbling over sideways, and start to roll down the incline; by quickly enrolling they soon reach the bottom of the hillock and then begin patiently to climb up again.

It would seem impossible that such a bulky creature with its specialised body design could climb trees, yet like the bright coloured *Chersastus* millipedes, at least two South African species are good climbers. *Sphaerotherium punctulatum,* one of the two large species of the genus, regularly climbs during the day, ascending vertical trunks or branches, clinging to slender twigs sometimes more than 3,5 m above ground. Ulrich Haacker reports that in 1972 he and a colleague collected 46 specimens of *punctulatum* by hand in 15 minutes at Illovo Beach, Natal; even by violent shaking it was impossible to dislodge them from the branches. This is all the more surprising seeing that the ratio of body length to width in nearly all species of *Sphaerotherium* is about two to one and the body width of the two largest species of the genus, *S.punctulatum* and *S.giganteum* is 21 and 32 mm respectively (approximately 1 inch), while the branches on which they walked were often quite slender vertical twigs. In the Knysna forest the same observer collected *Sphaerotherium cinctellum* almost exclusively on trees, usually the Red alder, *Cunonia capensis;* tree climbing obviously provides new and different food resources and climbing individuals of *S.punctulatum* were seen feeding on soft bark and green leaves.

The food of pill-millipedes consists of both the decayed and living parts of plants which includes morsels of fresh lichen and moss but the strong mandibles are able to deal with much coarser vegetable roughage such as tough leaves and pieces of bark; a certain amount of gritty sand is also swallowed with these items which assists in the detrition of the food. The intestine is longer than in most Diplopoda and bent on itself in the shape of an N, doubtless as a means of storing enough vegetable food material to nourish the large body.

## 7. *Enemy predators*

*Sphaerotherium* with its coat of armour is secure against the attacks of almost any predator, this and its enrolling reflex however are its only defences. Although the Glomeridia of northern Europe secrete a colourless sticky liquid which serves as a repellant, such a weapon is completely absent in *Sphaero-*

113

*therium.* At Cornell University, Thomas Eisner and his team of research workers carrying out tests on the defensive behaviour of millipedes, offered specimens of *Sphaerotherium punctulatum* and *S.giganteum* to several suppositive predators. The relevant parts of his interesting results read as follows:

In most cases the millipedes proved invulnerable as expected. Nevertheless they did fall prey to one particular enemy which was singularly adapted to cope with them. Ants, a blue jay and a grasshopper mouse were all equally unsuccessful. The blue jay pecked repeatedly at the millipede but its bill merely glanced off the hard shell of the prey, flipping it aside. The mouse, a voracious insectivore capable of subduing cockroaches of nearly its own size, seized the millipede with its front paws and attempted to bite it, clamping its jaws on the smooth shelled sphere; the prey was eventually abandoned uninjured.

The unexpected occurred in tests with the South African banded mongoose, *Mungos mungo.* The predator responded instantly to the glomerid, sniffing it and rolling it about with its paws. It seized it in its jaws, biting upon it with its sharp teeth, but the millipede was neither pierced nor crushed. Suddenly the millipede was dropped from its jaws and grasped by the front paws. The mongoose backed against a rock ledge in the cage, assumed a partially erect stance and − with a motion so quick as to be barely perceptible − hurled the millipede backwards between its legs (Figure 35). Fatally injured, with its shell broken and its body torn apart, the millipede was promptly eaten.

All told nine specimens of *Sphaerotherium* were offered to the mongoose. The results were virtually identical in every instance; the mongoose was inconsistent in its choice of target surface but it invariably selected an appropriate hard background and orientated itself properly forward of it just before the throw; sometimes the millipede was not smashed until the second or third attempt. The mongoose has a diversified diet and also eats eggs . . . it seems likely that glomerids are thrown and smashed in nature as they are in the cage.

## 8. *Distribution*

The Order, consisting of about 500 species, is divided into two suborders, the Glomeridia, which are mainly Palaearctic in distribution, and the Sphaerotheria which are African and Indo-Australian. The genus *Sphaerotherium* embraces all except one of the South African species, at least 50 in number, varying from about 10 to 58 mm in length. The forms living at high altitudes are always small while the large species are all found near the coast. These millipedes are for the most part confined to the dune forests of the coastal strip and to those parts of South Africa where fairly extensive inland forests occur, such as the Drakensberg foothills in Natal and the eastern and northern

Transvaal. They are almost completely absent from arid or semi-arid regions such as the Kalahari or most parts of South West Africa. Few of the individual species have a very wide distribution.

Figure 35. Three consecutive stages in the hurling and smashing of the pill-millipede *Sphaerotherium giganteum* by the banded mongoose, *Mungos mungo* (after Thomas Eisner and Joseph A.Davis 1969).

115

## E. THE COLOBOGNATHA OR SUCKING MILLIPEDES

It has been said that many small arthropods seem to spend their whole lives doing nothing and this appears to be true of the many small cryptic invertebrates that live in the humus of forests and especially of the Colobognatha, at any rate during the day-light hours. When uncovered beneath a piece of decaying wood they remain entirely immobile, and if, after some time they decide to go and find another refuge, they do so almost in slow motion. Such considerations apply of course only to those forms which have been observed in South Africa.

### 1. *General appearance and structure*

The Colobognatha resemble the wormlike Juliformia in their general shape and appearance and as their small legs are almost invisible, they are sometimes mistaken for worms; they are usually elongate and more slender than juliform millipedes and, as a rule, considerably smaller (Figure 36). The number of body segments is extremely variable; there may be more than 100, as in *Nematozonium* from Natal, or only 20 in the small enrolling form *Cylichnogaster,* with a body length fluctuating between 50 and 3 mm (Figure 37).

Figure 36b. The head seen from above (left) and from the side (right)

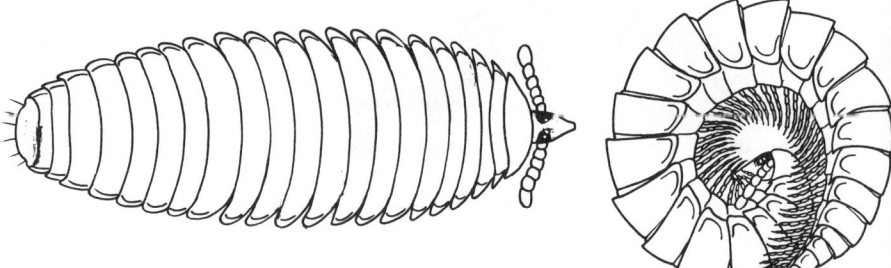

Figure 37. *Cylichnogaster,* a small conglobating Colobognath found only in indigenous forest of the Cape Peninsula.

116

In a number of anatomical details however they differ remarkably. The small head is drawn out into a characteristic elongated snout or beak; the mouthparts are much reduced and would not be able to deal with the rough woody material which is generally the food of the juliform millipedes; the mandibles especially are greatly reduced, being rod or needle-like for piercing rather than biting and entirely lacking strong teeth; they are capable of only a scraping action, being too weak to tear off large isolated particles of material. The lower plate of the mouth, the gnathochilarium, is also much modified and in general the weak construction of the mouthparts, and associated with them some strong muscles which can assist by their pumping action, point to a sucking method of feeding rather than a biting or chewing one.

Eyes are often entirely absent in this order and when present usually consist of one or two ocelli ringed with black pigment; they are never arranged in large groups composed of regular rows of ocelli as in the juliform milli-pedes (Figure 36). The antennae, large in proportion to the head, consist of the usual eight segments which are short and thick-set, the apical one with the four sense cones found in all Diplopoda.

The body segments are more often smooth and shiny, but in one of our species of *Nematozonium* they are covered with numerous fine short hairs giving them a soft, furry appearance. In cross section the body rings are not cylindrical like most Juliformia but hemispherical, somewhat flattened and wider than deep, the upper surfaces slightly arched. Nearly all the segments have defensive glands which are smaller than in the juliform millipedes and secrete a colourless liquid which has been said by some observers to smell of green wallnuts, by others of camphor.

## 2. *Habits*

Very little is known about the life cycle of South African Colobognatha which all live in damp dark corners and are not easy to discover in their retreats under rotting wood or in the moist humus of forests. The food, composed entirely of plant materials, is presumably taken in a decayed and perhaps almost fluid condition. The well known cleaning habits of other Diplopoda have not been observed and are apparently absent in the Colobognatha perhaps on account of the poorly developed mouthparts which may be unsuitable for this purpose, and of the greater rigidity of the body. This lack of flexibility has also had the effect of slowing down the walking movements of the group; the repellant secretions of the defensive glands and the spiral reflex seem to be their only means of defence.

Like the Polydesmoidea these millipedes with their relatively narrow heads use them as a wedge to enter a crevice which can then be further enlarged by repeated pushing movements of the body. Dr S.Manton found that with their strong body muscles and the ability of the ring segments to slide or telescope into each other they could exert more force than a Juliform millipede of the same size. One of the smallest South African species, like its counterpart *Siphonophora* in Asia and South America, is unique in being able to enroll, emulating the feats of the pill-millipede *Sphaerotherium*. In the case of *Cylichnogaster* however, while the legs and underside are well protected by the dorsal shields, enrollment is not so complete or efficient as in *Sphaerotherium* but represents a stage somewhere between the global enrollment of the latter and the ordinary clock-spring spiral of the wormlike millipedes (Figure 37). The little *Cylichnogaster,* about 3 mm in length, has been found in indigenous forest at Chapmans Peak on Table Mountain and nowhere else. It has another peculiarity for, unlike its relatives of the order, it is gregarious; adults of both sexes usually congregate in clusters of 50-100 under pieces of decaying wood.

## 3. Reproduction and development

The gonopods of the male are situated on the seventh and eighth segments but that on the eighth is leglike and not modified for the transfer of sperm. The gonopods present considerable uniformity throughout the various species; they are simple in structure being much less specialised and complex than those of almost all other Diplopoda.

The eggs, of which 65 have been counted in one nest, are laid in the spring. The female makes a small chamber of humus for them and curls herself round the egg mass, covering it with her flattened body and holding the loosely cohering eggs with her legs. If disturbed the female may desert the nest and in this event only a minority of eggs hatch out. In North America the female of *Brachycybe lecontei* transfers a clutch of eggs to the male who obediently incubates them by coiling himself around them.

During the breeding season the defensive glands of males and females in the vicinity of the nest are especially active and it is supposed by Verhoeff that the irritant effects of the secretion keeps enemies away from the eggs. The young leave the egg with four pairs of legs thus differing from the emergent larvae of all other Diplopoda which have three pairs; from time to time more pairs are added as they grow older. On hatching the young larvae speedily disappear into the ground. A number of larvae of *Burenia nasuta* captured together with adults by the writer in Natal, had seven or eight body segments and six pairs of legs. They were colourless and about 5 mm in length but in the size and arrangement of the setae and in the pigmentation of the eyes they resembled the adults. These doubtless represent one of the numerous stages between the first larva and the adult.

## 4. *Numbers and distribution*

The Colobognatha forms a restricted group representing a small fraction of the world millipede fauna. Although the approximately 270 known species have been found in various parts of the world, the continental faunas, apart from that of Europe, still remain comparatively unexplored. The first Colobognath from South Africa was described by C.Attems of Vienna as recently as 1928 and at the present time only four genera and about a dozen species are known to inhabit our region. These however are of exceptional interest and include one with the appropriate name of *Nematozonium longissimum* which probably holds a world record for the number of legs in any myriapod, namely 355 pairs.

## F. THE PAUROPODA OR DWARF MYRIAPODA

The extremely minute and primitive Pauropoda were first discovered in 1866 by John Lubbock who regarded them as a connecting link between the Chilopoda and Diplopoda. It is not surprising that they escaped notice for so long as they seldom exceed a millimetre in length, the largest known form being only 1.9 mm. They differ from all other myriapods in the peculiar branched antennae which resemble those of some Crustacea, such as the water flea *Daphnia* (Figure 38). Although forming part of the sub-phylum Myriapoda they are, like the Symphyla of the next section, neither centipedes nor millipedes but exist in their own right, occupying like them the rank of a separate class. In some respects, such as having the legs implanted at the sides, they resemble the centipedes, in some others the millipedes, but in most respects they differ from both.

### 1. *General appearance and structure*

The most striking structures of the head region are the prominent four-segmented antennae which divide at about the middle into two main branches. The upper of these segments carries one, the lower two long flagellae, each with numerous minute rings along their entire length which gives them the superficial appearance of being jointed. Between the two flagellae of the lower branch is a peculiar globular sense organ which has perhaps an auditory function; Verhoeff however regarded it as an organ of smell, comparing it with the four olfactory papillae at the apex of the antennae of Diplopoda. At the side of the head, behind each antenna, is another sense organ, the pseudoculus or ocular area, which is perhaps homologus with the pseudoculus of Protura and the organ of Tömösvary found in many other groups of myriopods, particularly the Oniscomorpha. The mouthparts are weak and very little specialised, consisting of an unsegmented pair of mandibles and a pair of maxillae more

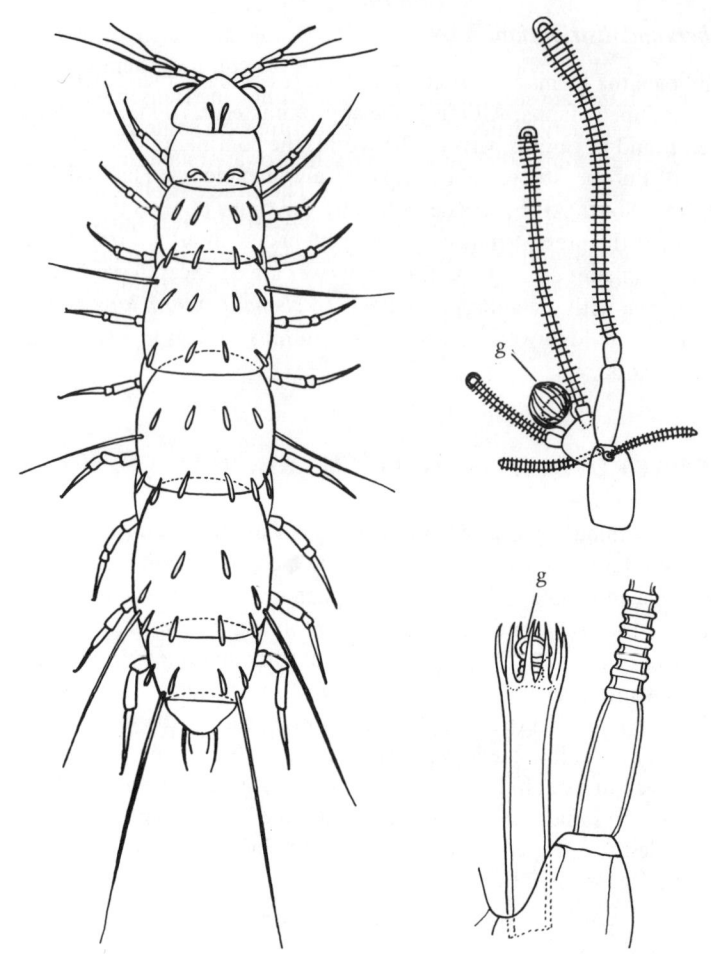

Figure 38. *Pauropus.* Left, the dwarf myriapod; right, the peculiar branched antennae in two different species with the globular sense organ (*g*).

or less concealed within the head capsule. They are more suited for sucking than for biting and resemble those of the Colobognatha, which are also adapted for this purpose.

The soft rather colourless body is usually fairly long and cylindrical though in *Eurypauropus* from Europe and the New World it is shorter and considerably flattened. It consists of 11 segments and a pygidium and is covered dorsally by six or ten simple tergites, the second to the sixth bearing a long tactile hair or trichobothrium at each side which stands far out from the body (Figure 38). These trichobothria are always provided with tiny accessory setae or prickles, which in some cases give them the appearance of

120

the minute plumule of a feather. There is considerable difference of opinion as to the function of the trichobothria; some authors consider them to be auditory organs but there seems more reason for following Hansen who believes them to be 'tactile hairs of specific structure with somewhat special functions'. Verhoeff thinks they represent a sense for apprehending changes of humidity and are sensitive to currents of air. The cuticle itself is usually soft and unchitinized, though in one genus, *Eurypauropus,* it is fairly strongly chitinized and covered with minute granules. The legs, nine in number, are primitive and simple, consisting of five segments ending in a double claw.

A respiratory system and a blood circulating system are completely absent and, as in small-bodied mites such as the Tyroglyphidae and many other excessively small animals, respiration can take place through the entire body surface. Organs or structures for defence are also lacking.

## 2. *Reproduction and development*

The sexes are separate, and in both the openings of the reproductive system are situated between the legs of the second pair; there is a well formed penis in the male and Schmidt thinks that true copulation must take place, though this has never been observed. The spermatophore, which consists of a ball of spermatozoa with a net-like covering, is attached to the ground by two threads.

Little is known of the postnatal development, but larval stages with 3, 5, 6 and 8 pairs of legs have been observed. Tiegs of Melbourne University, states that oviposition takes place in the early and middle summer months, the eggs being scattered singly in the decaying vegetation in which the animals live. They are white, spherical and of extremely small size, seldom measuring more than 0.11 mm (Fig 39a and b). The first larva with three pairs of legs has three well-developed tergites and one minute one, with two trichobothria on each side; at this stage it is a third to half a millimetre in length (Figure 39c). The five legged stage has an additional tergite and three trichobothria on each side (Figure 39d). The six legged stage has five tergites and four trichobothria on each side. The seven legged stage has five tergites and four trichobothria on each side, while the last larva with eight legs also has five tergites. New segments are formed between the last segment and those in front of it, development thus proceeding by anamorphosis. After moulting the eight-legged larva reaches the fully adult condition with its nine pairs of legs.

Harrison, the Australian naturalist, has described an instance of maternal care in the parent *Pauropus,* which, in one species observed by him, lays its eggs in groups of from 12 to 24 and mounts guard over them up to the time of hatching, after which the young are left to themselves; such behaviour on the part of the mother however is evidently an exception to the general rule, and has not been confirmed in the case of other species of *Pauropus.*

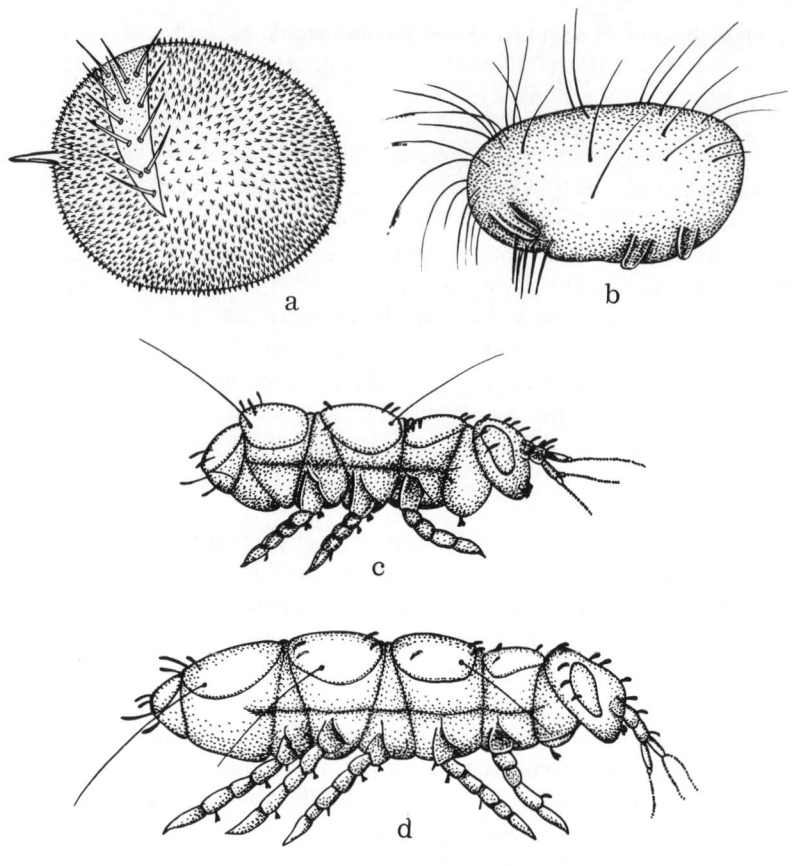

Figure 39. Development of *Pauropus*. The egg just after the first rupture of the shell a, b, followed by larval stages c, d.

## 3. *Habits*

Little is known of the general habits of these minute, secretive, and light-avoiding animals, living as they do in dark and hidden places such as fallen leaves in forest mould, crevices of damp wood or, as often in South Africa, in small depressions and fissures under stones, where on account of their minute size and neutral colouring, they are extremely difficult to see. They cannot long endure dry conditions and dessicate rapidly; even at room temperatures and humidities, their survival is usually a matter of minutes. They do not seem to be markedly gregarious or social, although an authority on the group, Professor Tiegs, has obtained several hundred specimens from a single damp log and successfully maintained large numbers of them in his laboratory for breeding purposes.

122

Dr Harrison writes about his captive pauropods as follows:

They are engaging little creatures to watch. They spend a great deal of time cleaning their appendages, commencing with the antennae, which are hauled down with a vigorous curling of the corresponding first leg about them, and drawn rapidly across the mouth, proceeding on to each pair of legs in turn, the limbs being systematically washed from coxa to tarsus. The attitude assumed, in cleaning the extended hinder limbs, is amusingly reminiscent of the same process in the domestic cat.

Thus not unlike the same process in our large Juliformia (p.83).

This particular species Dr Harrison observed was 'of a somewhat social habit, being almost invariably found in colonies, which vary from half a dozen to upwards of a hundred individuals'.

When their minute size is considered they are surprisingly agile; some like *Pauropus* are quite fast moving but others such as the European and north American *Eurypauropus* are sluggish.

The condition of the mouth-parts makes it improbable that the animal is able to feed on hard substances and these furthermore have not been found in the intestinal tract. It seems highly likely that its food consists of liquid or semi-liquid substances such as wood-sap or leaf mould in an advanced stage of decay. Verhoeff however is of the opinion that some of them at any rate do not refuse animal food. Harrison, speaking of Australian pauropods, is quite confident that they are humus feeders and Tiegs states that the intestines contain nothing but dissolved organic matter or the liquids which have been formed by the natural decay of damp particles of wood or leaves by which the animals are surrounded.

According to Tiegs they are preyed upon by the more slow moving pedipalps (Harvest-spiders?) and predaceous mites which form part of the associated microfauna of their environment.

## 4. *Distribution*

The Pauropoda have a world-wide distribution and the largest of the three families, the Pauropodidae, are found in all the continents of the world. Though perhaps not occurring in large numbers they are found in nearly all situations where damp soil, rich in decaying organic material, is to be found.

They can be looked for under stones and particularly in the humus of forests. The small number of species known, about 400, is entirely due to the neglect which the group has received at the hands of naturalists and collectors in general, and this again is due to their extremely small size and delicacy of structure. About 80 per cent of the species are attributed to the family Pauropodidae and about half of these to one genus, *Pauropus.*

They are not included in Attems' monograph of 1928 on the systematics of the South African Myriapoda as they were unknown in the subcontinent until 1930 when Herbert Womersley, the distinguished Australian entomologist, pointed them out to the writer when turning over stones on the slopes of Table Mountain. In 1955 the first specimen was captured by the writer in forest at Port St Johns and the first South African species was described as *Pauropus satelles* by Paul Rémy of the Natural History Museum, Paris. Two more were found on the foothills of the Drakensberg, Natal, at Champagne Castle and Cathedral Peak, localities about 6 000 feet above sea level. All the South African members are species of *Pauropus,* a genus found in all the countries of the world; one of them, *Pauropus huxleyi* var. *natalensis,* only differs in insignificant details from the European form.

## G. THE SYMPHYLA (Figures 40-42)

This primitive class of Myriapoda is a compact group consisting of a small number of forms which differ but little among themselves. As the name Symphyla implies, since its Greek derivations refer to a linking of different groups, they were considered to be a connecting link between the insects and Myriapoda. They can be looked upon as the survivors of a very ancient stock of myriapods related to the forms from which the insects were derived.

### 1. *General appearance and structure*

The Symphyla are small soft-bodied creatures, almost always dead white in colour (Figure 40). A large and full-grown specimen may perhaps be as long as 10 mm but most of them are considerably smaller. There are 15 tergites in most species but there may be as many as 24 and the legs are constant in number, consisting of 12 pairs, the first of which is either considerably shorter than the others or obsolete.

The head carries the antennae, consisting of a fairly large but variable number of small bead-like segments (22 in *Hanseniella*); they are fragile and very easily broken at any point along their length. The last segment has one to three sense organs at its apex which take the form of candelabra-shaped spines or hairs (Figure 41b); in addition there are rows of tactile hairs on all the segments. Eyes are absent but behind the antennae there is a sense organ of unknown function, the postantennal organ, which may be compared with the organ of Tömösvary.

The mouthparts consist of a pair of mandibles and two pairs of maxillae. The mandibles resemble those of Diplopoda in being two-segmented but are much weaker; with their comparatively simple structure they could hardly deal with anything but vegetable food materials. The rounded shield-like ter-

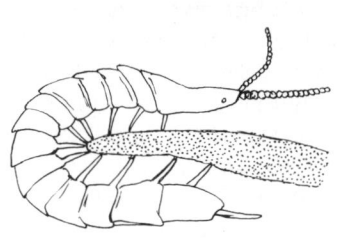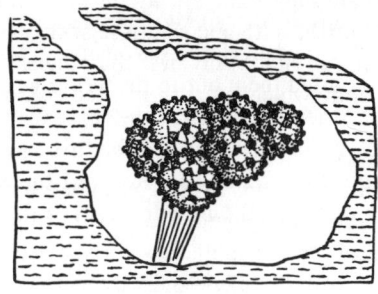

Figure 40b. Left, a young Symphylan bending its flexible body to climb round a leaf (after S.Manton), right, the underground brood chamber of another Symphylan, *Scutigerella,* with a cluster of eggs raised above ground on a stalk.

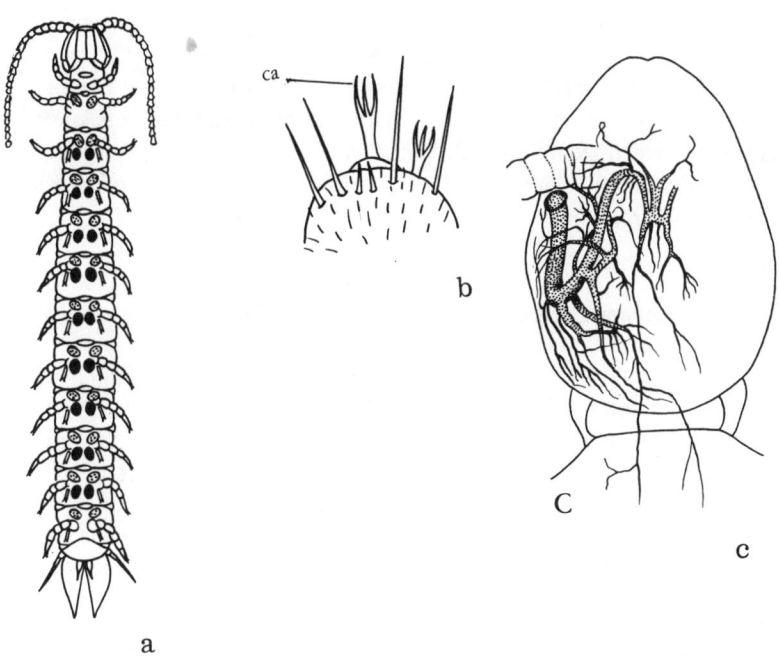

Figure 41. a, ventral surface of the Symphylan *Scutigerella* showing the coxal vesicles (black) and the posterior spinnerets; b, 'candelabra' olfactory organs on the antenna of *Hanseniella;* c, one side of the head of *Scutigerella* to show the single spiracle and greatly reduced tracheal system.

gites are simple and are always more in number than the leg pairs, as those of the fourth, sixth and eighth segments are duplicated; these extra tergites, which can slide easily over the others, give the animal the great flexibility of the body which is such a marked characteristic of this class of myriapods (Figure 40). The legs have five segments, the tarsus being equipped with two unequal claws. The coxae of the third to twelfth legs possess coxal vesicles, small sac-like organs which can be extended or withdrawn like the linings of a coat pocket (Figure 41a). At the posterior end of the body are two large conical cerci or spinnerets at the apices of which are the openings of the well-developed spinning glands. By means of these the animal is able to spin silky threads some of which are strong enough to support its weight and to be used for making a rough nest for itself. Between these cerci and the last pair of legs is a prominent tactile hair or trichobothrium at each side of the body.

The genitalia open between the fourth pair of legs on the fourth segment, in this respect resembling those of the Diplopoda. There is only one pair of spiracles, occupying a very unusual position, the head of the animal. These spiracles give off tracheal branches which supply only the head and two or three body segments (Figure 41c); it has been surmised that the thin-skinned coxal vesicles which seem to be filled with blood, partly serve the purposes of respiration in the remaining segments. A circulatory system comparatively simple in structure, is present.

## 2. *Egg-laying and development*

The sexes are separate and there are no secondary differences between the males and females.

The eggs, usually six to nine in number, though Filinger has counted clusters of 4-25 in *Scutigerella immaculata* and Tiegs 3-18 in *Hanseniella agilis,* are laid in the spring. They are globular, pearly white in colour, and about 0.5 mm in diameter, being covered with a series of anastomosing ridges which give them a roughened appearance (Figure 42a). The female deposits the eggs in a suitable cavity perhaps made by some other arthropod or in crevices of rotting wood, so that they are almost completely enclosed; in the enclosure the eggs, fastened together in a clump, are raised off the ground by a short stalk which prevents them from coming into contact with the walls or the floor of the brood chamber (Figure 40). The securing of the eggs to each other and to the substratum is probably effected with the secretion of the silk glands, and this isolation of the egg packet has the result of protecting it from fungal attacks.

An advanced embryonic stage has six pairs of legs when it escapes from the egg and it is then about a millimetre in length (Figure 42b); at first helpless and quiescent it becomes very active after a brief resting period. The young larva casts its skin after hatching; a whole series of moults then follows and with great regularity the larvae gain a pair of legs at each casting of the skin (Figure 42c); it follows that there are larval stages with 6, 7, 8, 9, 10 and 11

126

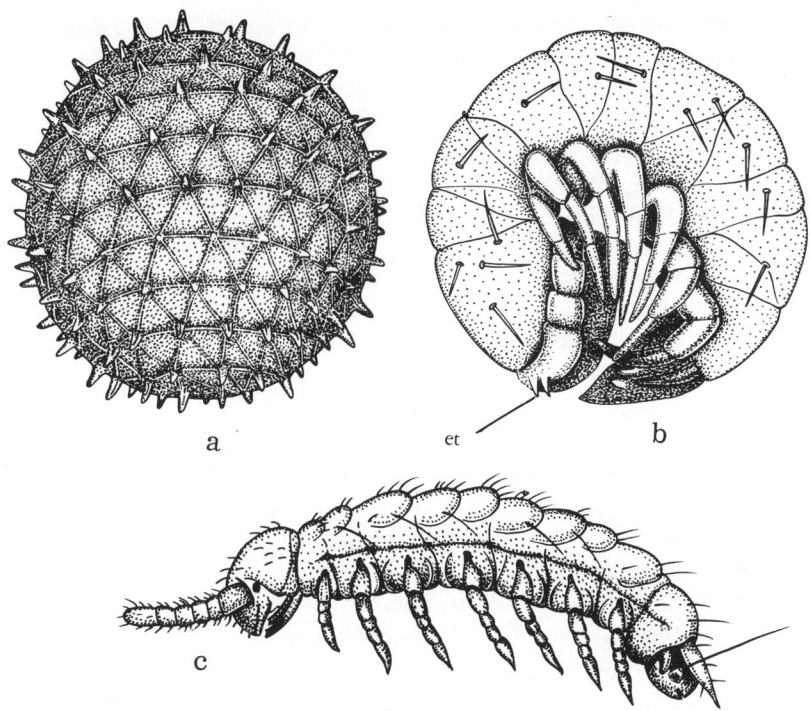

Figure 42. Developmental stages of the Symphylan *Hanseniella;* a, the egg; b, an advanced embryo shortly before emergence (*et,* egg tooth); c, first stage larva.

pairs of legs respectively before the final adult form, when 12 pairs of legs is reached. The growth period, from the leaving of the egg to the adult, requires about 50 days to complete.

Post-natal development is by anamorphosis, additional segments being interposed at a growing point between the last and penultimate segments. Filinger found that *Scutigerella immaculata* required 40-60 days to reach maturity in specimens from the United States.

The mother according to some authors, guards her eggs and the young in the same manner as some centipedes do theirs, but other observers such as Professor Tiegs, merely say that she guards them by waiting close beside them, as in the case of *Hanseniella.*

3. *Habits*

The Symphyla have little protection against loss of humidity and thus require an atmosphere with a uniformly high degree of moisture in their environment; they soon perish when exposed to dry air and the humidity of the soil is thus the most important factor in their economy. They are found in damp situa-

127

tions under stones or rotting wood, and occur predominantly in the humus layer of forests where they can be met with at all seasons of the year. They are sensitive to the slightest currents of air and when exposed to light by uncovering their hiding places, retreat as quickly as possible into crevices or under fragments of humus, though the delicacy of their structure makes it impossible for them to burrow in the ground for themselves. When walking the extremely flexible antennae are actively and restlessly engaged in sweeping the ground in front of them with lashing movements; in this way they doubtless explore the nature of the terrain by means of the numerous sensory spines on the underside of the segments.

Being extremely flexible and agile creatures, they can outwit pursuers by being able to make a sudden U-turn, or to change direction with a single quick movement; such tactics hinder pursuit by predators which cannot turn easily when in full flight. These little gymnasts can bend their supple bodies almost double in order to climb round an obstruction such as a leaf (Figure 40). They are able to use the threads spun from the silk glands of the spinnerets for letting themselves down from heights and so escaping into crevices. These threads are also used for making a rough web or nest and perhaps for securing the eggs to the sides of the brood chamber.

The food of the Symphyla is not known with certainty but most authors agree that it consists of vegetable matter such as various fungous growths, sporangia, and small fragments of decomposed wood, the diet thus resembling that of many of the small Collembola which live in a similar milieu. Filinger states that they feed on the small rootlets of young lettuces in the United States (*Scutigerella immaculata*). Tiegs says that the Australian *Hanseniella agilis* is 'undoubtedly omnivorous'. The structure of the mouth-parts would exclude the possibility of a purely carnivorous diet; they may however at certain times be scavengers as they have been known to feed on other dead Symphylids while arthropod remains have been found in the gut of *Hanseniella;* the intestine has also been found to contain large numbers of flagellate protozoa but these were probably taken in fortuitously with the semi-decayed food.

4. *The absence of defensive weapons*

These delicate and soft-bodied creatures have no structures for defence and probably rely on their agility, small size, retiring habits, and the threads of their silk glands for survival. They are not known to have any specific enemies and certainly do not prey on or pursue any other small arthropods, being quick to retreat when meeting with a foreign body or another animal which they may accidentally touch with the restlessly moving antennae.

The duration of life is surprisingly long for such small and delicate animals. Verhoeff kept adults of *Scutigerella immaculata* in captivity, and of eight specimens one lived for six years, and three for four years. The life cycle can thus be conservatively estimated as extending over six years. Moulting takes place frequently after the animals have become adult and probably continues throughout life; according to Michelbacher a single individual of *Scutigerella immaculata* may moult 40 times.

## 5. *Distribution*

Symphyla are found in all parts of the world, except the Arctic and Antarctic regions, and from sea level to considerable altitudes; in short wherever conditions are sufficiently humid and the soil contains a fairly high percentage of decaying organic material.

Collecting of the group has thus far been of a sporadic nature and only about 120 species are known. One of them, *Scutigerella immaculata,* has become widely distributed, probably by means of soil and plants, throughout Europe and the Americas. It seems to occur in far larger numbers than is usually the case with species of Symphyla, and is a genuine pest of a large range of cultivated vegetable crops. Attention has been drawn to the economic importance of this species by de Almeida in Portugal, and by Wymore, Filinger and others in the United States.

In South Africa *Hanseniella capensis* is widespread throughout the Republic but it never occurs in large numbers and is mainly confined to the soil of forests and plantations. It is seldom present in Museum collections and has not been observed to play any economic rôle. Up to 1938, *Hanseniella capensis,* first discovered by Hansen at Constantia near Cape Town, was regarded as the only representative of the Symphyla in South Africa, but Hilton in that year described another form from Natal, *Symphylella natala.* This species however is very inadequately described and figured and must for the meanwhile be regarded as doubtful.

# X. THE MILLIPEDES OF THE PAST

A number of fossil millipedes of fairly recent age are known to us from specimens which in past geological time were imprisoned and preserved in Baltic amber, the fossilised resinous droppings of coniferous trees. They are of Oligocene age and differ hardly at all from those forms living today; in the comparatively short time that has elapsed since they lived, a mere 20 million years or so, these conservative creatures seem to have changed very little.

The millipedes are derived from a very ancient family stem and were among the first animals to emerge from the waters and to walk upon the dry land. They are found as far back as the Devonian age and even in the Silurian age which preceded it, so that their ancestry dates back to a time considerably older than the Carboniferous period; when these early millipedes lived only a few primitive scorpions were able to exist on land while in the waters of the sea, corals, worms and starfishes flourished, together with the more ancient types of fish. But by the time that the coal measures were being laid down, some 250 million years ago, many more animals had been able to accustom themselves to life on dry land, though most of them had not advanced much further than the shores, swamps and lowlands of the continents of those early times; among such were some mites, ancestral spiders, scorpions and a number of millipedes.

The forests of the Carboniferous rose to their greatest heights after there had been a general lowering of the massive mountain systems of the preceding age by various denuding agencies. There was a general sinking and flattening of the land forming a series of shallow lakes and swamps that stretched endlessly to the horizon. The climate was hot and moist and the air strongly impregnated with carbon dioxide released by numerous active volcanoes; from the low and murky cloud covering crashing thunder storms and cloud bursts deluged the earth, feeding the lakes and swamps with incessant rain. Around the margins of these great areas of shallow water a luxuriant plant life lived and thrived; for the first time on earth primeval forests began to grow.

The millipedes are children of the forest; in them they have found security and ideal conditions of life as they still do in the giant forests of the Congo and the Amazon.

These primordial forests produced huge lycopods, club-mosses with tree-like forms such as *Sigillaria* and *Lepidodendron*. They did not however provide very much shade as these trees grew straight upwards with their narrow ribbon-like leaves gathered in a crown or fan at the summits, sometimes as far as 90 feet from the ground. It is not surprising therefore that in some of the North American coal measures fossil millipedes are often found in the boles of tree ferns such as *Sigillaria* where they had burrowed for shade and shelter, becoming fossilised together with the tree. On the other hand there was, as in our unspoilt modern forests, an ideal niche for millipedes in the luxuriant undergrowth of mosses, ferns and creepers which spread like a carpet beneath the larger tree-like forms.

These trees were of comparatively few kinds which repeated themselves in an endless succession over vast areas of land. Though green and tropical looking they would have seemed unendurably monotonous to our fastidious eyes, accustomed to the rich colours and contrasts of woodland and mountain slope, the changing flower patterns of meadow and prairie. The forest of the coal age, without the colours of flowers or the million tints of autumn, was a continual symphony of brown and green; there was no change of seasons then. When the wind blew, clouds of golden spore dust floated down, forming yellow veils on the black pools of the never ending marshes; this spore dust which was shed in such abundance was doubtless used as food by many insects, millipedes and other small creatures, just as it is today in the tropical rain forests of our modern world.

Mary Borden, in *Jehovah's Day,* has made us a lovely word picture of this primeval landscape – 'through the green gloom of the long vaulted colonnades of naked tree trunks, slowly, languidly, as snowflakes do, a pale golden shower of spores floated ceaselessly down from the leafy roof to the ground'. It was a soundless world too, for there was no singing of birds, no howling, roaring or squeaking of animals, large or small; no hum of bees or flies about the tops of the trees.

Over this wilderness of fleshy ferns and tree-sized club-mosses there brooded a deep heavy silence broken only by the thunder of tempestuous storms and shrieking gales or perhaps the melancholy crash of a falling tree fern or the splash of a webbed and clumsy-footed amphibian, to break the even flow of time.

The early millipedes of the coal age that lived in these silent swamps and forests were quite different from those living today. Though the various kinds of that time were recognisably like our present groups, such as the keeled millipedes, the pill-millipedes and the slender worm-like millipedes, they were in most cases much larger. Some of the giants, such as *Acantherpestes* for example (Figure 45), found in both North America and Europe, were nearly 2 feet in length and almost an inch thick; these monsters also had rows of curious forked spikes rising from their backs, looking like palisades of sharp stakes, and the use of these we can only guess at; perhaps they were a protection against the numerous small flesh-eating amphibia which haunted the swamps and forests of the coal age. It is possible that creatures like *Acantherpestes,* one of the worm-like Juliformia, also had gill-like organs which enabled them to breathe in water, and like those halfway animals, the frogs, were equally at home on land and in water. Perhaps some of them even had paddle-like limbs for swimming, for among the larger extinct millipedes some of the spines which protected their bodies could be moved (Figure 43).

The Carboniferous was an age of giants; just as the Mezozoic produced enormous armoured reptiles and later still the Cenozoic its giant mammals, in the Coal age it was the turn of the invertebrates. Large archaic dragonfly-like insects such as *Meganura* with a wing span of nearly 2½ feet or 76 cm flitted through the forest glades. The giant millipede *Acantherpestes* with its rows of spines was proportionately large. The scorpions were not to be left behind in the trend towards enlarged bulky bodies; *Gigantoscorpio* from the Carboniferous of Scotland was about 18 inches in length, while *Brontoscorpio* from an earlier epoch, the Silurian-Devonian, was more than twice this, a little more than three feet long. The largest of them all, a hitherto undescribed fossil, also from the Silurian-Devonian, is estimated to have been no less than six feet in total length.

The most striking of all is a myriapod-like fossil, a leviathan no less than 2 m, or more than six feet in length, by name *Arthropleura* (Figure 44). This creature had a superficial resemblance to a colossal millipede but cannot be placed in this class since it obviously has only one pair of legs to each segment; neither however is it a centipede for these legs were inserted close together in the same way as are the Diplopod pairs, nor can it be reconciled with any of the other myriapod orders that existed in the Coal-age period; in some res-pects it resembled those primordial arthropods, the Trilobites.

The bulky and clumsy body of *Arthropleura* proclaims that it was a poor walker on land; it was probably amphibious in habit and more at home in the shallow swamps which were filled with a rich compost of decaying woody substance on which it perhaps grazed. The thick-set heavy legs were strong but with very little flexibility, as the short segments were poorly articulated; they were however well provided with rake-like spines useful for collecting food as the undersurface of the body scraped along the bottom of the marshes, passing it forward to the mouth. The gut of the animal was found to contain fossilised plant fragments. In the same way the trilobites of the Cambrian period scavenged or fed upon small animals or dead organic matter dredged from the muddy bottom of the shallower seas.

Nearly all the fossil millipedes have been found in the northern hemisphere where they were probably far more widespread than is indicated by the few localities from which they have been recorded. Some of them appear to repre-sent the ancestral types, from which the Juliformia, Polydesmoidea and Onis-comorpha are derived and one, with tufts of needle-like hairs, has a superficial resemblance to *Polyxenus* and has been regarded as its hypothetical ancestor.

There were great earth movements and an uplifting of the continents to-wards the end of the era. An ice age overtook the flora of the southern hemis-

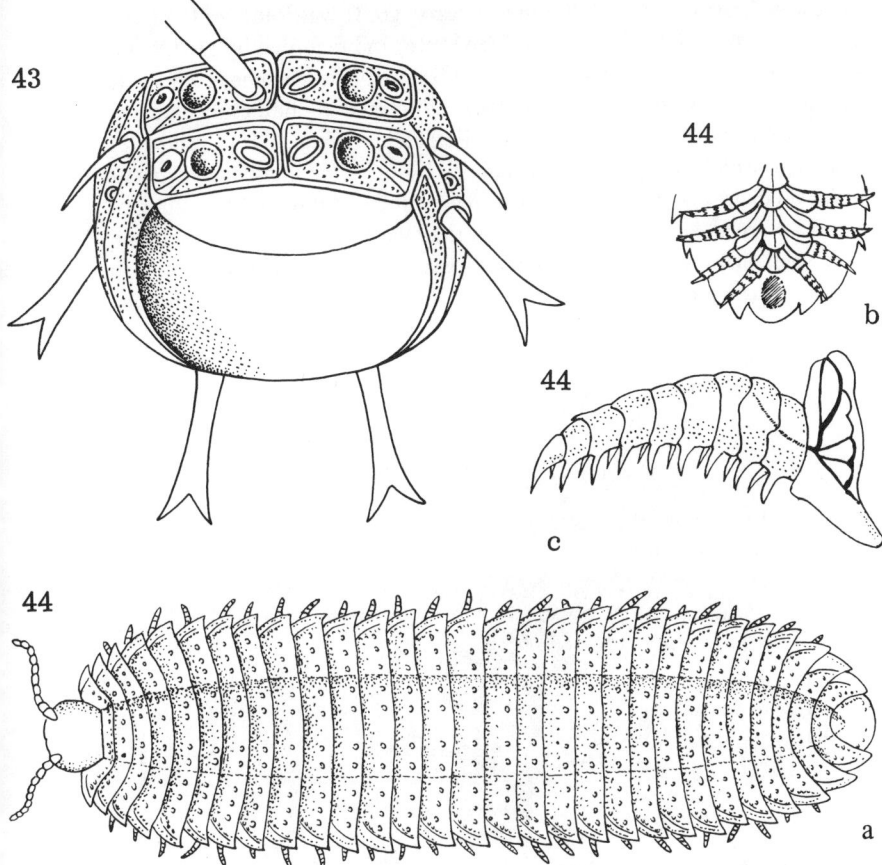

Figure 43. Above left, reconstructed body ring of the Carboniferous millipede *Acantherpestes*, the large forked spines of the back were perhaps movable.

Figure 44. a, reconstruction of an Arthropleuran, dorsal surface; b, a leg enlarged; c, under surface with the last four legs.

phere which had to cope with changed living conditions different from those of the contemporary northern flora. The great forests vanished, smaller and hardier plants took their place and the flora itself was more monotonous, characterised by such widespread plants as *Glossopteris* with its easily recognisable long tongue-shaped leaves.

This is no doubt why the millipede fossil fauna of South Africa is so meagre, to the point of being practically non-existent when compared with the large series of European and North American representatives of the same period. The only fossil millipede hitherto discovered in South Africa lived in the lower Triassic, 200 to 230 million years ago. It was found in the *Lystrosaurus* zone near Bethulie, Orange Free State, and recorded by Dr J.W.Kitching of the Bernard Price Institute, Johannesburg. The specimens appear to be juvenile, worm-like (juliform) millipedes contained in nodules of hard sedimentary rock; others were found near Bergville in Natal. These millipedes may have been gregarious in habit judging by the large number congregated in the small area of the nodule which may represent the fossilised dung of some large reptilian herbivore in which they were breeding and developing.

Nevertheless, there must have been a vast growth of plant life round the margins of the great central Karroo basin which existed like an inland sea

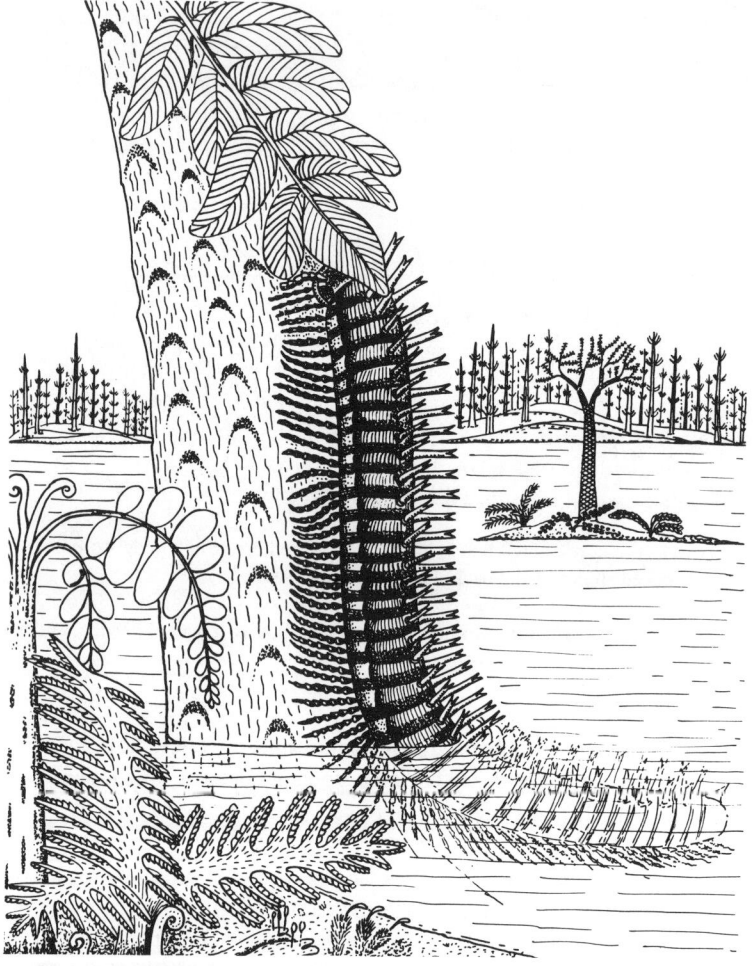

Figure 45. An imaginary scene from the Carboniferous age: the giant millipede *Acantherpestes major* (after S.H.Scudder 1882).

134

during this period; this vegetation supported the reptilian herbivores which lived there and which in their turn served as the prey of the carnivorous ones. A considerable fauna of millipedes, if not of centipedes, should have been able to live in this forest environment and any week or month now the press may announce to the world, as it did the Taungs skull and Coelacanth, that millipede fossils have been found in one of the wide-spread coal bearing strata of the Ecca series in Natal, Transvaal or Orange Free State. *Ex Africa semper aliquid novi.*

In anticipation of that day let us imagine that we have been suddenly transported by Mr Wells' Time Machine 250 million years into the past and are standing shortly after sunrise near the margin of one of the carboniferous swamps of the northern hemisphere. We may then see the giant millipede, *Acantherpestes,* about 20 inches in length and much more bulky than our largest thousand-legs, rearing itself from the water and climbing over the moss grown bank. The glossy rings shine in the rays of the rising sun as it moves smoothly and unperturbed towards a tree growing near, pausing now and again to browse on the damp and green moss that covers the slopes of the marsh; drops of water which cling to the curious spiky appendages along its back are glinting like jewels as they catch the light of the sun. It reaches the base of a large tree-fern and commences to climb, its body curved in a graceful arch, the legs moving upwards in rhythmic waves as the claws take hold on the rough scaly bark of the fern; it glides upwards until it comes to rest in the shady hollow at the base of the tuft of broad-leaved fronds which rise from the sodden wood and mould that has accumulated in this natural hollow; as the sun rises further it passes into an immovable trance or sleep, resting contentedly until the heat of the steamy tropical day will have passed or perhaps a brief thunder shower will have tempted it to move once more.

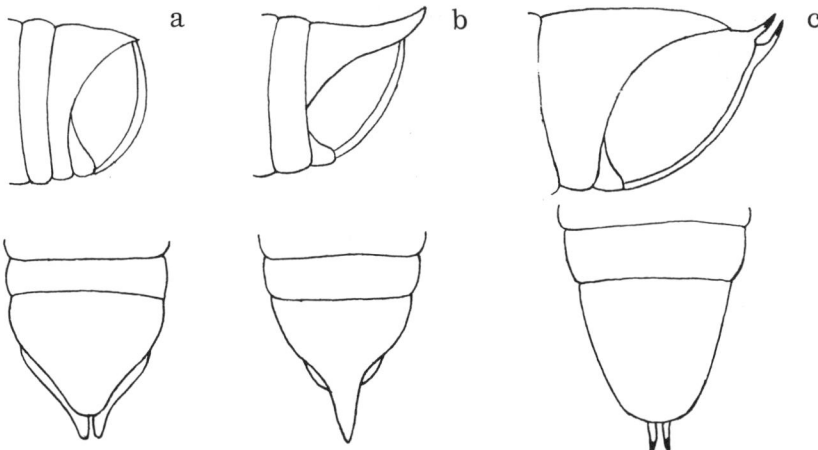

Figure 46. Above, the two last segments of a spirostreptid or a spirobolid millipede (a), a harpagophorid (b), and an odontopygid (c). Below, a dorsal view of the same segments.

(For text see page 138. left right)

# XI. THE IDENTIFICATION OF MILLIPEDES

The classification of nearly all millipedes presents a very difficult problem for even the Museum specialist so that making keys by which the layman can distinguish the individual species is beyond the bounds of practical possibility; the distinguishing characters which are mainly used are the highly complicated and elaborate sex organs of the males; female specimens are so alike outwardly that if we have to deal with them only the identification of species would in almost all cases be a matter of guess work; when dispatching millipedes to a specialist for identification therefore, it is essential to include at least one male specimen in the sample.

The five great divisions or orders of the Diplopoda can be easily distinguished from each other without special knowledge. A brief summing up of their main features, using only outward physical structures which can be clearly and easily seen when handling specimens, can be expressed as follows:

1. *Pselaphognatha.* 'Pin-cushion' millipedes, also known as *Penicillata*. These, being very much smaller than the other four groups, cannot be mistaken for any of them. In addition they have the characteristic tufts or bunches of modified hairs on each side of the body. Ocelli present in all South African forms. Four genera, with five or six species.

2. *Juliformia.* Worm-like millipedes, sometimes also called *Helminthomorpha*. They are elongate and cylindrical with 40-70 segments, the number varying in the same species. Fairly large groups of compound facetted eyes. Defensive glands opening by small pores at the side of the body on all segments except the first six. Normally reacting by enrolling in a flat spiral. All the largest millipedes, such as the Spirostreptidae, belong here. About 150 species.

3. *Polydesmoidea.* Keeled or 'flat-backed' millipedes. Short diplopods with 20 body segments, these constant in number. Body segments more or less hemispherical in cross section with lateral keels which when large and projecting give them a flat-backed appearance; in some genera these keels may be small and inconspicuous but when large are often fretted, sculptured or studded with rows of little tubercles forming characteristic, often elegant patterns. No eyes. Defensive glands large and well-developed but only seven in number, not arranged in regular sequence. About 150 species.

4. *Oniscomorpha.* Pill-millipedes. Short strongly built ovoid millipedes, the width of the body more than half its length. The eyes resembling those of the Juliformia. The body modified for complete conglobation, the body shields

very large, strong and tightly fitting, the last one, the pygidium, fitting over and concealing the head. Slow and clumsy walkers, the legs extremely flattened. No defensive glands. The last pair of legs in the males with a stridulatory apparatus for sound production. One genus *Sphaerotherium*, with about 50 species.

5. *Colobognatha*. Sucking millipedes. Obscure millipedes of sluggish slow-moving habits. Some species very long and slender, others quite short with few segments, resembling small pill-millipedes. Head drawn out in front into a characteristic pointed beak or snout. One or two large ocelli or none, when present surrounded by a conspicuous ring of black pigment. Defensive glands present on the body segments as in the Juliformia. The mouth-parts weak, greatly reduced, probably only suitable for a diet of soft or semi-liquid vegetable matter.

About five species of these uncommon millipedes are known. *Nematozonium* from the Drakensberg in Natal is blind, while the remaining genera have a pair of small eyes on each side of the head. *Cylichnogaster* from Table Mountain is extremely small and rather resembles a minute *Sphaerotherium*, being able to enroll in a ball. The other two, *Burenia* from Knysna and *Rhynchomecogaster* from the Drakensberg, Natal, are elongated worm-like forms.

In all about 400 species of millipedes are recognised in our country, although many more wait to be discovered and described. By far the largest of these five orders are the Juliformia and Polydesmoidea each totalling at least 150 species.

It is quite impossible to give keys for the numerous species or even for the genera and families of the worm-like millipedes (Juliformia); only a brief key has been attempted here in order to distinguish the three large subdivisions of the group which are presented in South Africa. Two of these have only a single genus, each with a limited number of species, all of which are small or of moderate size. The remaining subdivision contains the rest of the South African worm-like millipedes including all the largest forms.

The three subdivisions can be distinguished in the following manner:

1. The last joint of the legs of the male with a fleshy pad below; colour of the body usually a brilliant red, with a pattern of black or yellow spots . . . . . . . . . . . . . . . . . . . . . . . . . . . . . . Spiroboloidea (one genus, *Chersastus*)
— The last joint of the leg of the male never padded, the fourth and fifth joints however usually with a fleshy pad below; colour of body nearly always a uniform dull black, brown, yellow or greyish . . . . . . . . . . . 2
2. The first pair of legs of the male modified, often reduced to a rudiment; body size small . . . . . . . . . . . . . . . . . . . . . Cambaloidea (one genus, *Julomorpha*)

137

— The first pair of legs of the male not modified, resembling the remaining legs, usually large-sized millipedes . . . . . . . . . .   Spirostreptoidea

(See fig.46 on page 135)                                                                    (numerous genera)

The ecological milieu of *Chersastus* and its bright colouring has already been referred to. *Julomorpha* is a small grey millipede found only in the humus of the coastal forests from the Cape to the Transkei. The Spirostreptoidea are again divided into two further groups, the Odontopygidea and Spirostreptidea. In the Odontopygidea each of the two anal valves (the lateral semicircular discs at each side of the tail) are tipped with a small tooth on the upper surface, while the Spirostreptidea have quite smooth anal valves. The Odontopygidea furthermore are usually smaller and much more slender in form than the Spirostreptidea which contain all the giant millipedes; the response of an odontopygid millipede to handling is far more vigorous than in the case of a spirostreptid millipede and when caught it reacts energetically with snake-like writhings and twistings; often it will turn over and glide for some distance on its back. The large Spirostreptidea are more passive when handled and nearly always react to stimuli by rolling up in the form of a clock-spring spiral.

The largest of the odontopygid genera is *Spinotarsus* with almost 100 species which seems to have successfully occupied a larger extent of territory than those Juliformia which are confined to forests. It is found in a number of habitats with different conditions of humidity such as the open semi-arid lands while also adapting itself to grass prairie and thornveld as well as plantations of exotic trees. Finally it has reached and occupied the coastal forest in the extreme south of the continent as well.

The large group of Spirostreptidea can be further subdivided. It contains a small family, the Harpagophoridae, with only two South African genera which can be easily distinguished from other Spirostreptidea by having a large spike or tooth on the penultimate segment. Another family, the Trachystreptidae, living in Rhodesia, north-eastern Transvaal and along the east coast of tropical Africa, is also small, consisting of two genera, *Lophostreptus* with four species, *Calostreptus* with only one. They are easily recognised by their strongly striated body rings which, in contrast to the smooth exteriors of all other juliform millipedes, are covered with numerous, strong, more or less parallel longitudinal ridges, giving them a rather roughened appearance.

A fairly obvious difference in the outward appearance of the three millipede orders, Spirobolidea, Spirostreptidea and Odontopygidea is the slenderness of their bodies or otherwise. The Spirobolidea (*Chersastus* species), have a ratio of body length to body width of 7-9.5; in the Spirostreptidea this is 12, while in the Odontopygidea, by far the most slender of the juliform millipedes, it lies between 14 and 17. This applies to females; in males the disparity for all three groups is a little greater.

138

# XII. A SELECTION
# OF THE LITERATURE ON MILLIPEDES

Basic monographs, books and articles on millipedes in general.

Attems, C. 1909. Myriapoda. Schultze's Forschungsreise in West und Centralen Südafrika. *Denkschrift med. u. naturwiss. Ges. Jena* 14.

Attems, C. 1922. Myriapoda, in *Beiträge zur Kenntniss Land und Süsswasser faunas, S.W.Afrika.* Bd.2, Lief 1.

Attems, C. 1930. Diplopoda, in *Kükenthal's Handbuch der Zoologie,* 4.

Attems, C. 1928. The Myriapoda of South Africa. *Ann. S.Afr. Mus.* 26.

Attems, C. 1934. The Myriapoda of Natal. *Ann. Natal Mus.* 7.

Demange, J.M. 1981. Les Milles-pattes. Boubée, Paris.

Kästner, A. 1968. Invertebrate Zoology 2. University of Munich.

Lawrence, R.F. 1953. *The biology of the cryptic fauna of forests.* A.A.Balkema, Cape Town.

Manton, S.M. 1977. *The Arthropoda.* Oxford University Press.

Pocock, R.I. 1901. Millipedes, in *British Encyclopaedia,* 11th edition.

Verhoeff, K.W. 1926-1932. Diplopoda, in *Bronn's Klas u. Ord. Tierreichs.* 5.

## JULIFORMIA

The bibliography of this order is so enormous that it would be extremely difficult to make a useful selection from it and the reader should therefore consult the list of literature on millipedes in general (books and monographs) given above and the extensive bibliographies provided by these. A short list however is given of some minor items which may be useful together with the systematic works on South African Juliformia that have appeared since Attems' monograph, The Myriapoda of South Africa, in 1928.

Attems, C. 1914. Afrikanische Spirostreptidae, *Zoologica* 25.

Colville, F.H. 1913. The formation of leaf mould. *J. Wash. Acad. Sci.* 3.

Gerhardt, U. 1933. Zür Funktion de Gonopoden bei *Graphidostreptus gigas. Mitt. Zool. Mus. Berlin* 19.

Lawrence, R.F. 1938. Transvaal Museum Expedition to S.W.Africa and Namaqualand. Myriapoda. *Ann. Transv. Mus.* 19(3).

Lawrence, R.F. 1939. A new mite attached to the sex organs of South African millipedes. *Trans. R.Soc. S.Afr.* 27(3).

Lawrence, R.F. 1939. Notes on the habits of two mites living on South African millipedes. *Trans. R.Soc. S.Afr.* 27(3).

Lawrence, R.F. 1952. The unequal distribution of some invertebrate animals in South Africa. *S.Afr. Jour. Sci.* 48.

Lawrence, R.F. 1965. New Spirostreptidae and Harpagophoridae from southern Africa. *Mem. Inst. Invest. cient. Mocambique.* A17.

Lawrence, R.F. 1967. The Spirobolidae of the eastern half of southern Africa. *Ann. Natal Mus.* 18(2).

Lawrence, R.F. 1973. A revision of the Tachystreptidae (Diplopoda) of southern Africa. *Arnoldia* 6(4).

Pawlowsky, E.N. 1928. Section on Diplopoda, in *Giftiere und ihre Giftigkeit,* Leipzig.

Schubart, O. 1966. Diplopoda III. *S.Afr. Anim. Life* 12.

Silvestri, F. 1932. Istinti materni di alcuni Chilognathi. *Atti. Soc. Ital. Progr. Sci.* 3.

## POLYDESMOIDEA

In his paper on the keeled millipedes, Polydesmoidea, an American observer H.H.Miley has given an interesting account of the construction of the nest and the process of egg-laying in an American example of this order of millipedes. There is also an excellent series of photographs in J.A.Thomson's *New Natural History* illustrating the whole process from start to finish in the case of the European species, *Polydesmus complanatus.* Systematic work on the South African Polydesmoidea has been carried out by Attems, Verhoeff, Lawrence and Schubart.

Attems, C. 1928. The Myriapoda of South Africa. *Ann. S.Afr. Mus.* 26.

Attems, C. 1934. The Myriapoda of Natal. *Ann. Natal Mus.* 7.

Eisner, T. & H.E.Eisner 1965. Mystery of a millipede. *Natural History* 74(3).

Lawrence, R.F. 1953. A revision of the Polydesmoidea of Natal and Zululand. *Ann. Natal Mus.* 12(3).

Lawrence, R.F. 1958. Contributions to the myriapod fauna of Natal and Zululand. *Ann. Natal Mus.* 14(2).

Lawrence, R.F. 1959. A collection of Arachnida and Myriapoda from the Transvaal Museum. *Ann. Transv. Mus.* 23(4).

Lawrence, R.F. 1962. New Polydesmoidea from South Africa. *Ann. Natal Mus.* 15(14).

Lawrence, R.F. 1963. New Myriapoda from southern Africa. *Ann. Natal Mus.* 15(23).

Lawrence, R.F. 1966. The Myriapoda of the Kruger National Park. *Zool. Africana* 2(2).

Lawrence, R.F. 1970. Four new forest millipedes from Lesotho and the East Cape. *Ann. Cape Prov. Mus.* 8(5).

Miley, H.H. 1927. Life history studies of a millipede. *Ohio Jour. Sci.* 27.
Schubart, O. 1956, Proterospermophora. *S.Afr. Anim. Life* 3.
Verhoeff, K.W. 1939. Über Südafrikanische Polydesmoideen. *Ann. Natal Mus.* 8.
Verhoeff, K.W. 1939. Polydesmoideen, Colobognatha und Geophilomorpha aus Südafrika. *Ann. Natal Mus.* 7.

## THE PSELAPHOGNATHA (PENICILLATA)

This group has been inadequately dealt with in most of the natural history series of standard works. The few papers which have been devoted to the Pselaphognatha are scattered throughout the literature. Fortunately all this information has been incorporated in the general accounts given by Verhoeff and Attems. Verhoeff deals with the Diplopoda in general but those parts of his monograph which apply to the group can be obtained under various chapters and subheadings; his treatment of the order takes us to 1926 and a supplementary bibliography to 1932. For further reading the extensive bibliographies given by both these authorities should be consulted.

Pocock has given a very readable general account of the order in the 11th edition of the *Encyclopaedia Brittanica*. The section of his article devoted to the Pselaphognatha should be useful, if a general rather than a detailed description is desired. The post embryonic development of the group has been dealt with by Condé.

Hector has given some good photographs of the nests and the long crooklike bristles of the anal tufts in a species of *Polyxenus* from New Zealand. The South African species have been described by Attems and Condé.

Attems, C. 1936. Diplopoda, in *Kükenthal's Handbuch der Zoologie* Bd.4.
Attems, C. 1928. The Myriapoda of South Africa. *Ann. S.Afr. Mus.* 26.
Condé, B. 1949. Un Polyxenide inedit du Natal. *Bull. Soc. Ent. France* 1949.
Condé, B. 1959. Diplopoda Penicillata. *S.Afr. Anim. Life* 6.
Condé, B. 1962. Development postembryonnaire comparé des Penicillates. *Bull. Mus. Hist. nat.* (2e) 34(3).
Hector, C.M. 1935. Note on the occurrence in New Zealand of the Myriopod *Polyxenus. Trans. Proc. R.Soc. N.Zealand* 64, pt.3.
Pocock, R.I. 1901. Millipedes. *Encyclopaedia Brittanica*, 11th edition. Vol.18.
Verhoeff, K.W. 1926-1932. Diplopoda, in *Bronn's Klass. und Ord. Tierreichs* 5.

## ONISCOMORPHA – PILL-MILLIPEDES

To this order of millipedes Bourne has contributed an extremely useful paper on the anatomy of *Sphaerotherium,* the South African representative of the

141

group. Silvestri, and Attems in his monograph on the South African myriapod fauna, are chiefly responsible for the classification and description of the members of the order.

Bourne, G.C. 1885. The anatomy of *Sphaerotherium. J. Linn. Soc.* 19.

Eisner, J.C. & J.A.David 1967. Mongoose throwing and smashing millipedes. *Science* 155(3762).

Haacker, U. 1968. Das sexual Verhalten von *Sphaerotherium dorsale* (Myriapoda, Diplopoda). *Deutsch. Zool. Gesell. Innsbruck* 1968.

Haacker, U. 1972. Tree climbing in pill-millipedes. *Oecologia* (Berlin) 10.

Lawrence, R.F. 1966. The Myriapoda of the Kruger National Park. *Zool. Africana* 2(2).

Schubart, O. 1958. Diplopoda, Oniscomorpha. *S.Afr. Anim. Life* 5.

Silvestri, F. 1909. Materiali per una revisione dei Diplopoda Oniscomorpha. *Boll. Lab. Zool. Gen. Agraria, Portici* 4.

## THE COLOBOGNATHA

For general accounts of the group the monographs of Verhoeff and Attems, dealing with the Diplopoda as a whole, are the only ones available; Verhoeff especially has given a large amount of detailed information. Pocock's account though shorter is also well worth study. The systematics of the South African species have been undertaken by Attems and Verhoeff.

Attems, C. 1928. The Myriapoda of South Africa. *Ann. S.Afr. Mus.* 26.

Attems, C. 1951. Revision systematique des Colobognatha. *Mem. Mus. Hist. nat., Paris* NS3.

Cook, C.F. 1928. Millipedes of the Order Colobognatha. *Proc. US Nat. Mus.* 72(18).

Pocock, R.I. 1901. Myriapoda. *Encyclopaedia Brittanica* 11th edition, Vol.13.

Silvestri, F. 1896. I Diplopodi, I, parte systematica. *Ann. Mus. Civ. Genova* (2), 16.

Verhoeff, K.W. 1937. New Colobognatha from South Africa. *Ann. S.Afr. Mus.* 32.

Verhoeff, K.W. 1939. Polydesmoideen, Colobognathen und Geophilomorphen aus Südafrika. *Ann. Natal Mus.* 9.

## THE PAUROPODA

The anatomy of the group has been studied by Lubbock, Kenyon and other workers, its development and relationships by Tiegs. Attems and Verhoeff have in fairly recent years given general accounts of the class, that of Verhoeff being the most comprehensive. Of the more modern authors Manton's contribution in *The Arthopoda* cannot be missed. Remy and Hilton have added considerably to our general knowledge of the group while Hansen and Bagnall have given good systematic accounts. Paul Remy has described all of the few South African species.

Bagnall, R.S. 1935. A classification of Pauropods. *AMNH,* 16.

Hilton, W.A. 1930. The Distribution of *Pauropus. Science,* New York.

Hilton, W.A. 1930. Immature stages of *Pauropus. J. Ent., Zool. Claremont* 21, no.4.

Hansen, H.J. 1902. On the genera and species of the order Pauropoda. *Videnskab., Meddel., Naturh. Foren. Copenhagen.*

Hansen, H.J. 1917. On the Trichobothria in Arachnida, Myriapoda and Insecta. *Ent. Tidskr.* 38.

Harrison, L. 1914. On some Pauropoda from N.S.Wales. *Proc. Linn. Soc. N.S. Wales,* 39.

Kenyon, F.C. 1895. The morphology and classification of the Pauropoda. *Tufts College Studies* no.4.

Lubbock, J. 1867. On *Pauropus,* a new type of Centipede. *Trans. Linn. Soc.,* 26.

Remy, P.A. 1955. Description de deux nouveaux *Pauropus* d'Afrique du Sud. *Bull. du Museum* (2e), 27(4).

Remy, P.A. 1957. Palpigrades et Pauropodes du Natal (Recoltés du Dr R.F. Lawrence). *Bull. du Museum* (2e), 29(3).

Remy, P.A. 1959. Palpigrades et Pauropodes du Natal (Nouvelles recoltés du Dr R.F.Lawrence). *Bull. du Museum* (2e), 31(3).

Tiegs, O.W. 1947. The development and affinities of the Pauropoda based on a study of *P.silvaticus. Quart. Jour. Micr. Sci.,* 88.

## THE SYMPHYLA

Good general accounts of the Symphyla have been given by Attems and Verhoeff, that of Verhoeff being by far the most comprehensive of anything that has been written on the group. The morphology has been dealt with by Grassi, Haase and Schmidt, and the best classification of the group we owe to Hansen. Attems, Hansen and Hilton have recorded what little is known about the class in South Africa. Many papers on the habits and development of the Symphyla have been contributed by various authors, among the more recent being Williams, Muir and Kershaw, Wymore, Filinger, Remy and Friedel; Filinger has given a considerable list of references in his monographs on *S.immaculata.* Tiegs is the authority on the affinities and embryology of the Symphyla with his two monographs of 1940 and 1945. The following is a selection from the fairly extensive literature.

Attems, C. 1928. The Myriapoda of South Africa. *Ann. S.Afr. Mus.* 26.

Almeida, E.S.de 1930. Note sobre a *Scutigerella immaculata.* Newp. *Arq. Sec. Biol. Parasit. Coimbra* I, fasc.2.

Filinger, G.F. 1931. The garden Symphylid, *Scutigerella immaculata. Bull. Ohio Agric. Exp. Sta.* N.486.

Friedel, H. 1928. Oekologische und physiologische Untersuchungen an *Scutigerella immaculata. Z.Morph. Ökol. Berlin* 10.

Grassi, G. 1885. Morphologia della *Scolopendrella. Atti. Acad. Torino* 21.

Grassi, G. 1886. Morphologia della *Scolopendrella. Mem. Acad. Sci. Torino* 37.

Haase, E. 1884. Das Respirationsystem der Symphyla und Chilopoda. *Zool. Beitr. von A.Schneider,* Bd.I.

Haase, E. 1889. Die Abdominalanhänge der Insekten mit Berücksichtigung der Myriapoden. *Morph. Jahrb.* 15.

Hansen, H.J. 1903. The genera and species of the Symphyla. *Quart. Jour. Micr. Sci.* 47.

Hilton, W.A. 1938. *Symphylella natala. J. Ent. Zool. Claremont Cal.* 30.

Muir, F. & J.Kershaw. 1909. On the eggs and instars of *Scutigerella* sp. *Quart. Jour. Micr. Sci.* 53.

Michelbacher, A.E. 1938. 'Hilgardia' (Californian Agric. Exper. Sta.) 2.

Remy, P. 1936. La ponte et les premières stades larvaires des Symphyles. *Ann. Sci. nat. Paris,* 19.

Schmidt, P. 1895. Beiträge zur Kenntnis der niederen Myriapoden. *Zeitschr. f.wiss. Zool.* 59.

Tiegs, O.W. 1940. Embryology and affinities of the Symphyla based on a study of *Hanseniella agilis. Quart. Jour. Micr. Sci.* 82.

Tiegs, O.W. 1945. Postembryonic development of *Hanseniella agilis* (Symphyla). *Quart. Jour. Micr. Sci.* 85.

Verhoeff, K.W. 1933. Symphyla. *Bronn's Klass. und Ord. Tierreichs* 5.

Williams, S.R. 1907. Habits and structure of *Scutigerella immaculata. Proc. Boston Soc. Nat. Hist.* 33, no.9.

Wymore, F.H. 1931. The garden Centipede. *Bull. Univ. Cal. Agric. Exp. Sta. Berkeley,* no.518.

144

# INDEX